D0256812

THE FIRST OF EVERYTHING

Also by Stewart Ross:
Solve it Like Sherlock

Michael O'Mara Books Limited

First published in Great Britain in 2019 by
Michael O'Mara Books Limited
9 Lion Yard
Tremadoc Road
London SW4 7NQ

A CIP catalogue record for this book is available from the British Library.

Papers used by Michael O'Mara Books Limited are natural, recyclable products made from wood grown in sustainable forests. The manufacturing processes conform to the environmental regulations of the country of origin.

ISBN: 978-1-78929-062-2 in hardback print format
ISBN: 978-1-78929-209-1 in paperback print format
ISBN: 978-1-78929-063-9 in ebook format

1 2 3 4 5 6 7 8 9 10

Cover design by www.patrickknowlesdesign.com
Cover images © Shutterstock
Typeset by Design23

Printed and bound by CPI Group (UK) Ltd, Croydon, CR0 4YY

www.mombooks.com

To Lucy, without whose unstinting and generous efforts as proofreader, fact checker and warm-hearted supporter, this book could not possibly have been completed.

CONTENTS

INTRODUCTION

To research the first instances of literally everything (even the first book of firsts!) would be an infinite task. Inevitably, therefore, this book is selective. The criteria for inclusion are twofold. One, the very first of a *type* – for example, I have written about the first washing machines, including the first electric ones, but not about the vast subspecies of electrical washing machine, such as fully automatic, twin tub, etc. This should make plain the second criterion: I have included only such firsts as would, in my judgement, interest the general reader – who I hope is not a washing machine aficionado.

What makes this book different from other books of firsts? As well as being extremely broad-ranging and comprehensive, it is, as far as I am aware, the only one to put historical achievements in their rightful place. In other words, instead of concentrating on modern and largely Western gadgetry, I have attempted to give due weight to the inventiveness of our distant ancestors in the ancient civilizations of Egypt, China and the Middle East. In doing so, I have been surprised by how many supposed inventions of the industrial world (e.g. air conditioning) are in fact reinventions of or improvements on creations and behaviours of many thousands of years ago. By redressing the global imbalance that arises when we esteem modern technology over traditional ingenuity, we find the USA and the ancient world share the gold medal position on the podium of firsts, with the UK and France one step down.

I suspect few readers will be steadfast enough to read through the entire book, cover to cover, and most will dip into it for entertainment or use it as a work of reference for pub quizzes or to settle family arguments. To make these tasks easier, the contents have been arranged under three types of heading: the seven sections (In the Beginning; At Home; Health and Medicine; Getting About; Science and Engineering; Peace and War; Culture and Sport) are each divided into topics, which in turn are subdivided into subjects.

Finally, accuracy. Sources often differ widely, and precise dates are frequently controversial: is the date for the first of a certain type of machine, for instance, when it was dreamed up, when it was patented, when the prototype was built, or when it went into production? With this in mind, I have done my best to be clear and accurate; even so, in places I am sure I have fallen short. I apologize unreservedly for any confusion and frustration (even anger!) these inadvertent slips may cause.

Stewart Ross

NOTE: COUNTRY OF ORIGIN

Where relevant or known, the place where something was first used, discovered or invented is identified by the name of the present-day country occupying that region. This was not necessarily the name current at the time, and nor did ancient boundaries coincide with modern ones. Thus, firsts from Persia are generally labelled as 'Iran', Anatolia as 'Turkey', Mesopotamia as 'Iraq', etc.

PART I:
IN THE BEGINNING

BIG BANG

The first first, more or less by definition, was the Big Bang of some 13.8 billion years ago that created time, the universe… and everything. Everything? Even whatever it was that went bang? Let's not go there ...

LIFE

The first **life on Earth** – a much easier concept – is thought to have emerged 4.28 billion years ago, when our young planet was celebrating its 26 millionth birthday. Scientists refer to this first 'life thing' (the simplest of microorganisms) as LUCA – the Last Universal Common Ancestor. Apparently, we're all descended from LUCA.

'HANDY MAN'

LUCA's descendants took a very long time to evolve into the **genus *Homo***, which appeared only about 2.1 million years ago. Its identifying features were an ape-like physiognomy, a bulging brain and an ability to use primitive **tools** (another first) – hence its name, ***Homo habilis*** or 'handy man'.

UPRIGHT, FIRE, TOOLS AND SPEECH

About 200,000 years later, ***Homo erectus*** ('upright man') had developed. This creature's even larger brain may have enabled it to speak (if so, we have the first **language**). The species had possibly learned to handle **fire** (one more first, see p.13) and certainly made tools of greater sophistication; it was also spreading out from Africa to populate the globe.

HOMO SAPIENS

We're not quite sure what happened next. However, from among the variety of *Homo* types, by *c.* 50,000 BC ***Homo sapiens*** ('intelligent man') was hunting and gathering in the inclement conditions of the last glacial period. These were the first modern human beings, forebears of the extraordinarily inquisitive and inventive men and women responsible for the astonishing catalogue of firsts that follows.

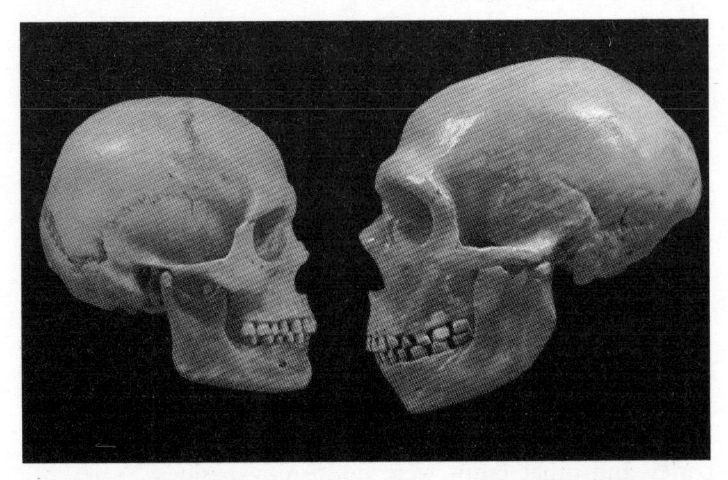

The skulls of *Homo sapiens* (left) and of his unsuccessful relative, *Homo neanderthalensis*

PART II: AT HOME

CIVILIZATION

EVOLUTION

Human development is **evolutionary and cumulative**; or, in the words of Sir Isaac Newton (repeating a well-known idea of classical origin), new ideas and objects came from people 'standing on the shoulders of giants'. We have already met some of those giants, the primitive, anonymous people who made three vital early breakthroughs: fire, tools and speech.

AGRICULTURE

The baton now passes to *Homo sapiens* for perhaps the most significant of all firsts: **agriculture**. The change from hunting animals and gathering food to considered husbandry took place independently in some dozen different parts of the world, beginning with the domestication of pigs in Mesopotamia (Iraq) in about 13,000 BC. Within a few thousand years – again in the Middle East – fields of wheat, barley and other crops were flourishing beside pigsties, all tended by the first **farmers**.

SETTLEMENTS AND CITIES

Once human beings ceased being nomadic, caves, tents and temporary shelters were abandoned in favour of **permanent dwellings**. It is not known if Jericho, in the Israeli-occupied West Bank and Çatalhöyük in Turkey (both *c.* 9000 BC) were the first towns – with populations of several hundred they were more like modern villages – but they are certainly the oldest to have survived to the present day.

HOMES

DOORS AND HINGES

A house requires an entrance. The earliest depiction of **doors** is in ancient Egyptian tomb paintings, and the first **hinges**, reserved for temples, tombs and palaces, were simple pivots set into the lintel and sill (doorstep). The working of bronze (from *c.* 3300 BC and later iron, see p.18) allowed sturdier hinges to be made, and by Roman times the device was considered sufficiently important to merit its own deity: Cardea, the goddess of the hinge! The modern **butt hinge**, set flat into the door and frame, does not appear until about 1850.

WINDOWS

The first **window** was just a hole in the wall to let in light and fresh air, and give smoke and smells a way out. When necessary, windows could be covered by a piece of wood, cloth or hide – the first **curtains**. By the first century AD the Chinese were manufacturing **paper** (see p.10), a material they used as a window covering as well as for writing. The first

A window in the ruins of Pompeii

glass was made in the Levant about 5,500 years ago, and by AD 100 the Romans were using glass fashioned in Alexandria as windowpanes. Though crude and relatively opaque, it was better than the sheets of thin translucent stone and flattened animal horn that remained in use for centuries. **Stained glass** dates back to ancient Egypt and Rome; its extensive use in windows began in the Christian era, when it was employed to stunning effect in the monastic buildings of northern Europe. **Plate glass** production had to wait another millennium, beginning in London, UK, during the reign of James I (1603–25), and becoming more widespread (e.g. King Louis XIV's palace at Versailles, France) after the pioneering work of Louis Lucas de Nehou and Abraham Thevart in 1688.

THE PAPER PROCESS

While the ancient Egyptians wrote on mashed reeds (**papyrus**), the Greeks and Romans on the skins of dead animals (**parchment**), and the Meso-Americans on prepared bark (**amate**), it was the Chinese who first made paper. The court official Cai Lun documented the process in AD 105, possibly because he reckoned his papermaking process was better than anyone else's. Paper was essentially different from (and superior to) the other three writing surfaces as the properties of the constituent fibres changed fundamentally during the manufacturing process.

Papermaking in ancient China

LOCKS AND KEYS

Once people had houses with doors and windows, and accumulated precious possessions within them, they needed **locks** to keep everything safe. Again, the Middle East led the way with locks made of wood. The first with metal components appeared in ancient Rome and China. The first **key** is attributed to Theodorus of Samos (sixth century BC), about the same time as the first **padlock** was made. **All-metal locks** are attributed to an unknown Englishman during the reign of Alfred the Great (AD 871–899) – possibly to safeguard against Viking pillage.

CHUBB AND YALE

The modern lock was made feasible by the precision engineering of the Industrial Revolution. Robert Barron (UK) manufactured the **lever tumbler lock** in 1778, and Jeremiah Chubb (UK) went one better in 1818 with a version that could be opened only with its own key. Then, in 1848, the American Linus Yale invented the modern **double-acting pin tumbler lock**, opened with the familiar Yale key.

ELECTRONIC SECURITY

Security remained largely mechanical until the twentieth century. In 1909, in a sign of things to come, Walter Schlage (USA) devised a door lock that would also turn lights on and off. Five years later, the luxury Scripps-Booth automobile boasted the first **central locking** system. It was not until the 1970s, however, that the electronic security revolution really took off, beginning with Tor Sørnes's programmable electronic **key card** in 1975 (Norway). Thereafter, it was chips with everything, including **car keys** (1980, Ford, USA) and **passports** (1998, Malaysia).

LEGENDS OF A LOCKSMITH

Jeremiah Chubb's 'unpickable' lock of 1818 gave rise to two stories, both dubious! The first is that the device was adopted by Portsmouth Royal Dockyard (UK), where Chubb may have worked, after the Prince Regent accidentally sat on it. The second tells of a convicted burglar – a locksmith by trade – being offered a free pardon if he could unpick Chubb's device. After two months, he admitted defeat – and was returned to the prison hulk in Portsmouth harbour from whence he had been taken.

FIRE AND COOKING

THE HEARTH

Using fire is very different from *making* it. Hominoids appear to have discovered the secret of **fire-raising**, probably by friction between hard and soft wood, between 300,000 and 100,000 years ago. To date, the earliest **hearth** we know of is a 300,000-year-old example in the Qesem Cave near Tel Aviv.

CONTROL OF FIRE – A MAJOR FIRST

Fire was present from the beginning of the Earth; indeed, one can say it was the beginning of the Earth. Early *Homo* species first learned to control it and use it for their benefit between about 600,000 and 300,000 years ago. This slow process – part deliberate, part opportunistic – was one of their greatest achievements, arguably as significant as their development of agriculture. Fire provided warmth, encouraging them to move into inhospitable regions; a fire in the mouth of a cave or on the edge of an encampment provided protection against wild beasts; fire extended the range of activities available on cold, dark evenings (for example whittling, chipping and storytelling); fire broadened culture (enter the charcoal artist and sculptor of fired clay figures); and, above all, fire produced cooking: the first (and possibly brain-enhancing – see cookers, below) step on the long and tasty road from mammoth steak to Michelin star.

COOKERS

Cooking was widespread by 100,000 BC. The process did not just cheer up the taste buds. One school of thought (the 'Cooking Hypothesis') believes cooking food enabled our brains to develop into the current 50+ terabytes version by providing more brain food and cutting required feeding time.

Simple **roasting** over an open fire expanded into **pit ovens** (29,000 BC), front-loading **bread ovens** (*c.* 800 BC, ancient Greece), hand-turned **iron spits** (medieval), purpose-built **brick and tile ovens** (fifteenth century, France), **iron stoves** (*c.* 1720, Germany), iron **ranges** (burning wood and coal, *c.* 1800, UK), to **gas ovens** (1826, James Sharp, UK) and finally **electric ovens** (*c.* 1890, Canada). The Scotsman Alan MacMasters invented an **electric toaster** in 1893, and General Electric (USA) put a model on sale in 1909; we had to wait another ten years before our breakfast slice popped up. Kleen Maid **Sliced Bread** (USA) for use in toasters went on sale in 1928. **Microwave cookers** became available in 1946, and the first **induction hobs** were sold in 1973 (both USA).

FIRES AND SMOKE

With an open fire virtually the only form of non-solar heating for thousands of years, it's surprising no one thought of the **chimney** before the twelfth century; Fontevraud Abbey, France, boasts the earliest extant examples. The Chinese manufactured the first smokeless fuel – **coke** – in the fourth century BC. The **iron fire back** dates from the fifteenth century, in Europe, the **cast-iron stove** from 1642 (Massachusetts, USA), and the **raised grate** (supposedly the brainwave of Prince Rupert, the dashing nephew of Britain's Charles I) from *c.* 1678. Stoves burning **anthracite** or coke (a by-product in the manufacture of coal gas) date from the 1830s, with the first **gas fires** available some twenty years after this.

CENTRAL HEATING

Some say central heating dates back seven thousand years to the Korean ***ondol***, meaning hot stone. However, most give credit to the Greeks and Romans for their **hypocaust** underfloor system of circulating hot air. Much later, in the sixteenth century, Hugh Plat (UK) dreamed of a **piped steam** greenhouse heating system, but the concept did not become a (short-lived) reality until the end of the eighteenth century. It was superseded by **piped hot water**, a method pioneered by Russia's Peter the Great for his St Petersburg Summer Palace (*c.* 1710). Russia also gave us the **radiator** (*c.* 1855). Since then, the main developments have been in ways of heating the water: the idea of a **heat pump** (1855–7) is said to have been the brainchild of the Austrian Peter von Rittinger. In 1896, US resident Clarence Kemp was supposedly the first to deliberately heat water by using **solar power** (he painted a large tank black); in 1948, Robert Webber (USA) invented the **ground source heat pump**.

... AND COOLING

The first record of an **icehouse** for the preservation of food was in the time of King Zimri-Lim (*c.* 1780 BC, Syria). Three thousand five hundred years later, the Scotsman William Cullen built the first machine to **produce ice** (1756). Practical continuous **refrigeration** capable of cooling beer in the outback did not arrive until the Scottish Australian James Harrison's machines came into operation in 1857. **Domestic fridges** went on sale in 1913 and **freezers** – made by General Electric (USA) – in 1927. Twin compartment **fridge-freezers** appeared in 1939 (USA).

COOL BUILDINGS

Evaporative cooling was the first method of deliberately lowering the temperature in a room. The ancient Egyptians built **windcatchers** (*c.* 3000 BC) that cooled the breeze by passing it across water; they also trickled water over wind-ruffled reeds hanging over a window. The Cool Hall in the palace of Emperor Xuanzong of China (*c.* AD 712–756) was apparently cooled by water-powered **fans**. The first electric **desert cooler** was patented in 1906 (USA). Four years earlier, Willis Carrier of Buffalo (USA) had installed the world's first **electric air conditioner**.

PRESERVING

Freezing is not the only way to preserve food. Farmers were **drying** crops in the sun at least 14,000 years ago. Not long afterwards, it was discovered that adding salt enhanced the process (something similar was used to **mummify** human bodies: 5500 BC in Chile, 3000 BC in Egypt). Alcohol's preservative qualities were noted when early **alcoholic drinks** were made in China and the Middle East, about 7000 BC. The first known **pickling** (cucumbers from the Tigris Valley) was in 2030 BC. **Canned** (tinned) foods were eaten by the Dutch navy from at least 1772 and the canning process was patented in 1810 (UK). The previous year, Frenchman Nicolas Appert had discovered how to feed Napoleon's armies while on the march by preserving cooked food in a jar. Mass-produced **frozen food** was introduced by Birds Eye in 1930; the first complete 'TV Brand **Frozen Dinner**' went on sale in America in 1953.

IN THE KITCHEN

THE FIRST COOKING POT

Simple roasting required nothing more than a piece of pointed stick, bone or antler to hold the food near the fire. Advantage: quick and easy – even a child could do it. Disadvantage: a lot of nutrients (especially fats) dripped off into the fire. Answer: the cooking pot. It's likely that animal shells were the first cooking receptacles. The first man-made pots did not coincide with the arrival of agriculture, as was once thought, but were fashioned perhaps 20,000 years ago by Chinese hunter-gatherers. And what did they prepare in those large, acorn-shaped ceramic vessels? The world's first fish soup!

POTS, PANS AND KITCHENWARE

Clay storage pots have been around for some 20,000 years, the earliest examples being Chinese. The ancient Egyptians came up with **glaze** about 10,000 years ago, and their Mesopotamian neighbours the **potter's wheel** some 4,000 years after this (see p.145). The Chinese created **porcelain** (hence 'china') around 1600 BC, although – surprisingly – the invention of **bone china** is credited to the Londoner Thomas Frye in 1748. Because it was rare and rustless, naturally occurring **gold** may well have been the first **metal** used by humans for jewellery, but weight and rarity left the gold cooking pot in the realm of fairy tale. **Copper**, first **smelted** in the region of Serbia *c.* 5250 BC, was fine for pots and pans, and **bronze** (fifth millennium BC) even better. The first bronze alloy comprised copper and **arsenic**, with obvious drawbacks, and was soon replaced by a **copper–tin** mix (4500 BC, Serbia region). **Brass**, a copper–zinc alloy, was first made in the Middle East during the third millennium BC. **Iron** smelting began in Turkey *c.* 2100 BC.

Copper, brass, iron and **steel** (see p.141) cookware remained the norm until the appearance of **enamelled** pots (late eighteenth century, Germany), and **aluminium** (first extracted in 1824) pans at the end of the nineteenth century. Long before this, the Italian Luigi Brugnatelli had invented **electroplating** (1805). The following century witnessed the arrival of **Pyrex** glass (1908, USA), **stainless steel** (1913, UK), **Teflon** (1938, USA), **carbon fibre** (1860, Joseph Swan, UK), and a number of other newfangled composites, including modified ceramics and glass.

THE LARDER

BREAD

It has been suggested that our hunter-gatherer forebears varied their pick 'n' mix diet by making some sort of **bread** (grains mushed up with water and heated on a stone) about 30,000 years ago. In 2018, evidence was found in Jordan of bread **baked** 14,000 years ago. The ancient Egyptians were **leavening** bread with yeast by the first millennium BC. Around 800 BC, the Mesopotamian invention of the **millstone** produced a finer flour (Iraq). **Baking** was made a profession in ancient Rome (*c.* 300 BC, Italy), and with Rohwedder's **automatic bread slicer** (1928, USA) we arrive, more or less, where we are today.

LET THEM EAT CAKE

Baking leads us to **cakes**, where again the ancient Egyptians claim a first with their honey-flavoured confections. The earliest published recipe for a **layered cake** is American (1872). The first **pudding** is impossible to identify because the word embraces both sweet and savoury dishes, including **pies**, supposedly invented by the ancient Greeks (fifth century BC, referred to in the plays of Aristophanes). This means they also had **pastry**, first devised in Egypt, Phoenicia or Greece in the first millennium BC.

THE SAUSAGE

The art of stuffing meat and other products into washed-out lengths of intestine to produce a **sausage** appears first in China and ancient Greece at about the same time (*c.* 600 BC), though some believe the Mesopotamians were doing something similar 2,000 years earlier. The fermented sausage – **salami** – was made in the Greco-Roman world at least two millennia ago. Shortly after the French Revolution (1789), the French wrapped **croissant** pastry (a seventeenth-century Hungarian invention) around a sausage to produce the **sausage roll**, which was almost immediately exported to the UK, where it became an institution.

HOT DOGS

Germans were eating Frankfurter Würstchen sausages as long ago as the thirteenth century. When they emigrated to the USA, they took their delicacy with them and served it on the streets as a snack. Problem: how to eat a piping hot sausage without getting one's fingers burned. Answer: provide customers with returnable gloves. New problem: customers did not return the gloves and the sausage seller's profits collapsed. New answer: serve the sausage in a bread roll – and thus the hot dog was born. No one is sure when this was first done, but a good claim is that of a Mrs Feuchtwanger of St. Louis, Missouri, USA, in 1880.

DOMESTICATING ANIMALS

'Man's best friend' is also his oldest: the dog became the first **domesticated animal** about 15,000 years ago (Germany). Next came the **pig** (China) and the **sheep** (Anatolia), both around 9000 BC, with the **goat** (Persia) sharing fourth place (*c.* 8000 BC) with **cattle** (Western Asia). The **chicken** from India (*c.* 6000 BC) came some way behind.

DAIRY PRODUCTS

Butter was probably discovered by accident shortly after the domestication of the goat (see above). **Margarine** was patented by the Frenchman Hippolyte Mège-Mouriès in

1869. Unilever launched **Stork** in 1920 and plant-oil-based **Flora** in 1964 when health-conscious Scandinavians were beginning to use a variety of alternative spreads based on vegetable oils, olive oil and other products. **Cheese**, too, was discovered by accident when milk was stored in the rennet-rich stomachs of ruminants. Poland and Croatia provide us with the earliest evidence of deliberate **cheese-making** (*c.* 5500 BC), and we know the dates when certain types of cheese were first recorded: Cheddar (*c.* 1500, UK), mozzarella (1570, Italy), Parmesan (1597, Italy), and Camembert (1791, France). The first **cheese factory** was Swiss (1815), as was the first **processed cheese** (1911).

At some unspecified time following the domestication of the horse (see p. 103), the tribes of Central Asia are reputed to have accidentally discovered yoghurt – an idea supported by the Romans describing it as 'barbarian' food. Production was industrialized in 1919 by the Turk Isaac Carasso who named his business after his son, Daniel. Later, Danone – 'little Daniel' – was Americanized as Dannon. The commercial production of fruit and flavoured yoghurts began in the Czech capital, Prague, in 1933.

FAST FOOD

Any bite grabbed by a person on the move – a knight in the saddle, a peasant at the plough, or a shopper in the market – can be described as 'fast food'. Consequently, though the term did not enter dictionaries until the early 1950s, it is impossible to say when the first fast food went on sale. To keep it simple, we define it as pre-cooked, takeaway food served in an urban environment; and the pioneer accolade is shared between John Lees of Mossley, Lancashire, UK, and

Customers at the first MacDonald's, opened in 1955 in Illinois, USA

Jewish immigrant Joseph Malin of London's East End. Both were serving **fish and chips** in the 1860s.

Among the many claims for the invention of the **hamburger**, I favour that of Frank and Charles Menches and their butcher Andrew Klein, late of Hamburg, Germany, who ran out of pork for their sausage stall at the 1885 Erie County Fair, Buffalo, USA, and substituted spiced ground beef in a bun – with notable success. Walter Anderson's White Castle Diner (1921, Wichita, USA) is often described as the first **fast-food restaurant**. The first **McDonald's** opened twenty-seven years later. The **kebab** has its roots in medieval Persia, and there is written evidence of Italians feasting on **pizza** as long ago as AD 997.

Curried and sweet-and-sour dishes are as old as cooking itself in the Near and Far East. The first **Indian restaurant** outside India opened in 1810 (London) and London's first

Chinese restaurant in 1908. **Takeaway** Indian and Chinese dishes were offered from the 1950s onwards. **Balti**, an original Anglo-Indian dish, is said to have originated in Birmingham, UK, in 1971.

Apocryphal or not, the story of the notorious gambler John Montagu, 4th Earl of Sandwich (1718–1792), calling for food to eat while he was at the tables is as good an explanation of the first **sandwich** as any. **Baguettes** were defined in law in 1920, though production had started in eighteenth-century France when, presumably, they were first sliced open and filled with butter, cheese and ham. Shortly afterwards, William Kitchiner was telling the world (*The Cook's Oracle*, 1817, UK) how to make the king of nibbles – the **potato crisp** (or chip). Smith's Crisps (UK) were the first to include a twist of **salt**, and in the early 1950s Irishman Joe 'Spud' Murphy manufactured the first flavoured crisps.

BREAKFAST CEREALS

Porridge is as old as agriculture (see p.7). German-American Ferdinand Schumacher's German Mills American Oatmeal Company (1854, morphing into **Quaker Oats**, which was swallowed by PepsiCo in 2001) launched the USA's breakfast cereal revolution. On the other side of the Atlantic, the

Early Kellogg's advert for toasted corn flakes, 1907

production of **Scott's Porage Oats** (bought by Quaker, 1982) had to wait until 1880. The first **man-made breakfast cereal** was James Caleb Jackson's 'Granula' (1863, USA). Eighteen years later, John Harvey **Kellogg** (USA) produced something very similar and called it 'Granola' to avoid being sued. Kellogg's Battle Creek Toasted **Corn Flake** Company was founded in 1906, six years after Swiss doctor Maximilian Bircher-Benner had fed his patients with the first **muesli**.

NAUGHTY BUT NICE

Ice cream (of sorts) was enjoyed by ancient Greeks (fifth century BC), though this was little more than flavoured snow. Ice cream proper, using **milk**, was made in China three centuries later. **Flavoured ices** had reached **Europe** by 1533. Ice-cream **cones** made their appearance in the 1880s (Italy) and **soft ice cream** in 1934 (USA). Some 5,300 years ago, people living on the eastern slopes of the Andes were the first to cultivate **chocolate**. In the first half of the sixteenth century, the Spanish brought it to Europe where it was transformed from a drink into a **bar** by the Dutchman Coenraad van Houten (**pressed chocolate**, 1828) and the Englishman Joseph Fry (**mouldable chocolate**,

An Aztec with a chocolate pot and hand-carved whisk (sixteenth century)

1847). With Swiss Daniel Peter's invention of **milk chocolate** (1875), using the **powdered milk** (1867) of another Swiss, Henri Nestlé, the modern chocolate industry was up and running.

SWEETNESS

Sugar cane was first cultivated in India about 8000 BC and **granules** were made from its sweet juice around AD 350. In 1493, Columbus introduced sugar cane to the New World, and the first **sugar beet** processing factory opened in Silesia (Poland) in 1801. A third source of sweetness, glucose-rich **corn syrup**, was discovered in Russia in 1812 and made commercially in the USA in 1864. Its attraction for human food products was transformed when the Clinton Corn Processing Company (USA) managed to turn some of the glucose into **fructose** (*c.* 1958). Following further research in Japan, by 1967 Clinton was producing **HFCS** (high-fructose corn syrup). The sweet-tasting properties of **saccharin** (originally 'the poor man's sugar') were accidentally discovered by its creator, the German scientist Constantin Fahlberg, in 1879. Hermesetas (1932, Switzerland) was the first branded **artificial sweetener**.

DIETS

The figure and health conscious have always watched what they eat. In the second century AD, the Greek gynaecologist Soranus recommended weight loss through a regime that included **laxatives** and **purgatives**. (His name hints at the unpleasant side effects!) The first **modern diet** is said to be that proposed by Dr George Cheyne (UK) in his 1724 *Essay* *of Health and Long Life*, and the first popular **diet book** was written by one with plenty of experience of the effects of poor nutrition: undertaker William Banting's *Letter on Corpulence* (1863, UK) urged a balanced diet of meat, vegetables and fruit, washed down with a dry wine. It was still in print in the twenty-first century. The first **diet pills** – potentially fatal – went on sale in several Western countries towards the end of the nineteenth century. Far safer were the **Weight Watchers** meetings of the early 1960s.

THE CELLAR

WATER AND WINE

Water was our first drink – and it remains the world's favourite. No doubt our prehistoric ancestors livened it up with various juices and so forth, but the dramatic breakthrough came with **alcoholic drinks**. Archaeological evidence suggests the Chinese got there first when they fermented grapes some 9,000 years ago. Whether or not this was **wine** is a moot point. Less equivocal evidence points to Georgia as the first producer of wine, around 6000 BC, and Armenia as possessing the oldest **winery** (*c.* 4100 BC). Identifying wines by their **vintage** dates back to late medieval France. **Cork** stoppers, used as an alternative to wood and rags since Roman times, were made cheaper with the manufacture of **compound cork** (1890, Germany). The great cork vs synthetic debate, initiated in 1955 by the invention of the **plastic closure**, became even more fierce when the **screw cap** for wine was successfully trialled in Switzerland in 1970.

DANGEROUS FIZZ

The first sparkling wine was simply ordinary wine that bubbled because it had not finished fermenting – a phenomenon noted in the ancient world. The first reference (1531) to deliberately fizzy wine – Blanquette ('Little White') de Limoux – was made by the Benedictine monks of the Abbey of Saint-Hilaire who distributed it in flasks sealed with cork stoppers. Champagne came later, largely thanks to the demand of seventeenth-century English drinkers for more wine from the Reims region that continued to ferment (hence the fizz) in the bottle. The English were also responsible for bespoke champagne bottles. (The original ones were liable to explode, injuring cellar workers – who wore protective face masks – and occasionally destroyed 90 per cent of a year's production in a chain reaction.) Dan Gurney, American winner of the 1967 24 Hours of Le Mans, was the first driver to spray champagne in celebration of a motor-racing victory.

SPIRITS

Distillation began in Babylon (Iraq) 3,200 years ago, and by the first century AD it was being used in China and the Indus region (Pakistan) to produce alcoholic beverages. During the ninth century, the Arab scientist Al-Kindi was making a drink we would recognize as **spirits**. The earliest reference to **whisky** (1494) comes, unsurprisingly, from Scotland. The next century heralded the arrival of all the other great spirits: the Dutch found that burned wine (**brandy**) travelled better than unburned; by the time of their revolt against Spanish rule (at the end of the sixteenth century), their soldiers were swigging jenever (**gin**) to give themselves courage; at least fifty years before that, Polish merchants had introduced Russians to **vodka**. The first **cocktails** were served either in London (1798, UK) or the USA (1803) – whoever had the idea, they were onto a winner.

BEER

It is possible that beer was first brewed about 13,000 years ago in what is now Israel. Other archaeological finds suggest that around 8000 BC, the inhabitants of the Near East (Turkey) became the first brewers. More reliable evidence points to the amber nectar being first supped in Persia some 2,500 years after this. Its popularity soared rapidly, with workers on a variety of strenuous projects (including building the Pyramids of Giza) being rewarded with daily pints, quarts or even gallons!

TEA AND COFFEE

For those who like legends, the first **tea** is said to have been consumed in 2737 BC by Emperor Shennong of China when he sipped boiled water into which leaves of the tea plant

A seventeenth-century coffee house

(*Camellia sinensis*) had accidentally blown. Mythology aside, the pioneers of tea were certainly Chinese. Tea was taken to **Japan** around AD 800 and reached **Europe** in 1606. The earliest reference to tea in Britain is 1658. **Coffee** originated in Ethiopia. It was associated with Islamic practice and the first recorded consumption is in a fifteenth-century Yemeni monastery. The first European **coffee house** opened in Venice (1645).

FIZZY DRINKS

The development of fizzy drinks was made possible by the invention of **carbonated water** by the English chemist Joseph Priestley in 1767. By the 1770s, Thomas Henry of Manchester, UK, and the Swiss Johann Jacob Schweppe

Way past its sell-by date: a tin of roast veal from 1823

(who later moved to London, UK) were selling a product they named '**soda water**' to the public. Fizzy **lemonade** was available in the 1830s (USA and Europe), and **tonic water** from 1858 (India). The first effective gas-proof **crown bottle seal** (1892, USA) and **glass-blowing machine** (1899, USA) facilitated the industry's rapid expansion. Beer was the first drink to be **canned** (1933, USA), with soft drinks following the next year. **Aluminium cans** replaced steel in 1959, the same year as the **ring pull** (both USA).

COLA

The caffeine-rich (up to 4%) kola nut has been held in high esteem in central West Africa for millennia, but it was not exploited commercially in the West until the nineteenth century. **Coca-Cola**, invented by US pharmacist John Pemberton in 1886, was the first cola drink. Its great American rival was first sold by its creator, Caleb Bradham, as 'Brad's Drink' (1893 – the name was changed to **Pepsi-Cola** in 1898). Pepsi was the first to produce a **diet** version (1963).

SHOPPING

MARKETS AND SHOPS

Basic shopping – i.e. exchange in order to acquire – began when the first civilizations introduced specialized labour. Thus, farmers exchanged their produce for tools made by the blacksmith, and so forth. Since it was convenient for many such exchanges to happen in the same area, around 7000 BC **markets** grew up in places like Çatalhöyük (Turkey) and Kashan (Iran). **Money** was first used about 1,000 years later (see p.205). As marketplaces became established, manufacturers found it convenient to site their workplaces on the edge of the retail area, enabling them to sell directly to customers – and so we get the first **shops** (literally 'workshops'). The Chinese are credited with the first **branding** and product-specific **packaging** (*c.* 200 BC).

SHOPPING CENTRES

Numerous cities claim to possess the world's first **shopping centre**. The disagreements stem partly from the lack of clear distinction between an arcade, a suk, a bazaar, and a mall. **Trajan's Market** in ancient Rome (built AD 100–110) has its advocates, though **suks** in Aleppo and elsewhere in the Middle East and on the Silk Road (at least 600 years old) dispute this; others vote for La Rochelle's late medieval **arcades** (France) or Istanbul's **Grand Bazaar** (begun 1455, Turkey). St Petersburg's Gostiny Dvor (opened 1785, Russia) may well be the first **purpose-built shopping mall** and London's Burlington Arcade (1819, UK) the first **purpose-built shopping arcade**. The Southdale Center, the world's first fully enclosed, climate-controlled **shopping mall**, opened in Edina, Minnesota, USA, in 1956.

CO-OPS, CHAINS AND DEPARTMENT STORES

Founded in 1498, the Shore Porters Society of Aberdeen, Scotland, was supposedly the world's first **cooperative organization**. The earliest **retail consumer cooperative** was the Fenwick Weavers' Society (1761, also in Scotland). The first **cooperative bank** (1852, Germany) was established by Franz Hermann Schulze-Delitzsch. In 1792, Henry Walton Smith opened a shop that would become the world's first **chain of stores** (WH Smith, UK) – America's first chain, the Great Atlantic & Pacific Tea Company, was launched sixty-seven years later. The year 1796 saw the opening of two shops that lay claim to be the world's first **department store**: Harding, Howell & Company on Pall Mall, London, UK, and Watts Bazaar (becoming Kendal, Milne & Faulkner or simply 'Kendals') in Manchester, UK. Parisians reckon Le Tapis Rouge (1784), though technically a **novelty store**, may justly claim precedence over its British rivals.

SUPERMARKETS, HYPERMARKETS AND LOYALTY SCHEMES

Michael J. Cullen opened King Kullen, the world's first **supermarket**, in New York City, USA, on 4 August 1930. Europeans took the idea a step further when Carrefour opened the first **hypermarket** (supermarket and department store combined) near Paris, France, in 1963. **Trading stamps** as a reward to shoppers for their loyalty originated with Schuster's Department Store in Milwaukee, Wisconsin, USA (1891), while the **loyalty card** (1981) and the **Air Miles** scheme (1988) were both pioneered in the UK.

CHECKING OUT AND DRIVING THROUGH

Self-service supermarkets were all very well, but where were customers supposed to put all the stuff they planned to buy as they wandered around the aisles? Sylvan Goldman, owner of Oklahoma's Humpty Dumpty supermarket chain (USA), came up with the answer in 1937 – the **shopping trolley/cart**. The answer to one problem produced another: what to do when all those stuffed trolleys piled up at the checkout as harassed staff struggled to read the prices on each item? Enter the **barcode**, patented in 1952 and accepted as a practical commercial concept in 1966 (both USA). The first **drive-through bank** window became available in 1929 (USA), and the first **drive-through grocery collection point** – another American innovation – opened in California in about 1941.

Thus far, all these shopping innovations required checkout and payment. Then, in 2018, came Amazon Go in Seattle, USA, the first **checkout-free store** in which the cashier was replaced by cameras and sensors. Big Brother was watching.

CLOTHING

MATERIALS

Apparel was originally fashioned from animal skins (leather) and suitable vegetable matter (e.g. leaves). **Needles** (Siberia and South Africa) for sewing these materials together are about 50,000 years old, and the oldest **thread** (Georgia) was twisted together from strands of wild **flax** 34,000 years ago.

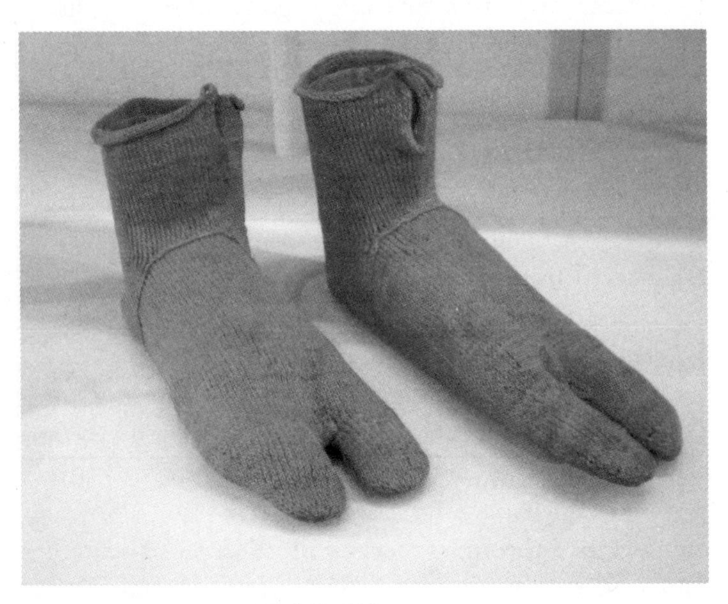

Ancient Egyptian socks, AD 300–500

Then came the process of turning animal hair, especially wool, into a **textile. Felt** came first (perhaps 8000 BC, Middle East) because it required no special technology. Next was **nålebinding** ('needle-binding', *c.* 6500 BC, Israel), and finally **spinning** and **weaving**, first practised in ancient Egypt some 5,400 years ago. The earliest **materials** were flax (made into linen cloth) in the Middle East, **silk** in China, **cotton** in India, **wool** in the Middle East and Europe, and **hemp** in China and Japan.

LOUSY EVIDENCE

We believe our humanoid ancestors began wearing clothes at about the same time as the exodus from Africa, over 100 millennia ago (anthropologists disagree on the dating). Evidence for this supposition comes not from the clothes themselves – none remain – but from the creatures that lived in them. No, not us, but our lice. Head lice, which attach themselves to body hair, have been with us for ever. But geneticists tell us that body lice, which lay their eggs in clothing, evolved from head lice only around 100,000 BC – giving us a date for when we started dressing up. This ties in with the African exodus, too – clothes became a necessity when we migrated to colder climes.

TECHNOLOGY

Spinning was initially done by hand (possibly 20,000 years ago), rolling the threads or twisting them into yarn with a spindle. India invented the **spinning wheel** between 1,000 and 1,500 years ago. The process was mechanized by James Hargreaves's **spinning jenny** (1764, UK) and Richard Arkwright and John Kay's **spinning frame** (1769, UK). Subsequent improvements included **rotor spinning** (1963, Czechoslovakia) and **friction spinning** (1973, Austria). Weaving derived from basket making. The creator of the **loom** is unknown, as different cultures appear to have constructed their own versions independently in early historic times. Less controversial is John Kay's mechanical **flying shuttle** (1733, UK), leading to the semi-automatic **power loom** (1785, Edmund Cartwright, UK). Frenchman Joseph Marie Jacquard's **punch card loom** (1804) enabled the automatic weaving of complex patterns, and in 1895 James Henry Northrop built a **fully automatic power loom** (USA). The **shuttle-less loom** was the work of the Sulzer Brothers (1942, Switzerland); **water-jet** and **air-jet looms** were developed in Czechoslovakia in the 1950s. Not to be left behind, knitting was mechanized as long ago as 1589 when British clergyman William Lee made the first **knitting machine**.

SYNTHETIC FIBRES

Today, at least half the world's clothing is made from materials unknown 150 years ago. Joseph Swan exhibited a laboratory-made **cellulose fibre** in 1885 (UK); Hilaire de Chardonnet created **artificial silk** in 1889 (France); Courtaulds (UK) produced **rayon** in 1905, and DuPont (USA) came up with **nylon** in the late 1930s. In 1941, John Whinfield and James Dickson of the British Calico Printers' Association Ltd invented **polyester** fibre ('Terylene'), perhaps the most successful artificial clothing material. Americans Robert Kasdan and Stanley Kornblum made it into the first **wicking microfibre** in 1998. The 1960s witnessed two dramatic developments. First came **Spandex** in 1962 (USA). Never heard of it? Try **Lycra**! Two years later, Stephanie Kwolek of DuPont (UK) came up with life-saving **Kevlar**, the ultra-strong material inside bulletproof vests (see p.44).

BASIC CLOTHING

The first (i.e. oldest extant) **skirt**, woven from Armenian straw, is dated around 3900 BC. A 5,000-year-old linen **shirt** (or perhaps a **dress**) was found in an Egyptian tomb. Not long after, **saris** were being worn by women of the Indus Valley. **Trousers** were not far behind. The first to survive, discovered in China, may be 3,000 years old. A written reference to trousers comes from the sixth century BC, when they were described as the preferred garb of Asian nomads who spent long hours on horseback. The Middle Eastern **thawb** predates the arrival of Islam and its creation may well be contemporaneous with the flowing robes favoured by ancient Greeks and Romans of both sexes (first millennium BC).

BOOTS AND SHOES

Fossilized toe bones from China suggest humans first wore **shoes** 40,000 years ago. The earliest surviving **sandal** is made of sagebrush (*c.* 8500 BC, Oregon, USA), and the oldest **leather shoe**, found in an Armenian cave, is 5,500 years old. The first picture of a **boot** is in a Spanish cave painting of *c.* 13000 BC, and boots themselves – already a symbol of power and authority – were worn in ancient Persia around 3000 BC. Wooden **clogs** were probably first worn by the Celts of northern Europe, perhaps in the early years of the first millennium AD. At the end of this period, Persian cavalrymen were wearing **high heels** – to help their feet stay in the stirrups; Shāh Abbās's emissaries introduced them (and waistcoats, see p.44) to Europe in the seventeenth century.

FOOTWEAR FASHION

Britain's Queen Victoria wore **elastic-sided** (Chelsea) boots in 1837; on the other side of the Atlantic, the horsemen of Kansas, USA, were inventing the **cowboy boot** by combining European height with Mexican heels. Transatlantic ingenuity went the other way when Anglo-American entrepreneur Hiram Hutchinson moved to France to make **rubber boots** ('wellies' in British English) in 1853. **Flip-flops**, based on the Japanese *zōri* sandal, emerged in 1950s USA. The Frenchman André Perugia, whose highly fashionable shoes were available

High heels at the court of Louis XIV of France, 1728

from 1906 onwards, may have created the **stiletto heel**, though the term 'stiletto' when applied to shoes appeared in print only in 1959. Thick-soled **brothel creepers** had made their silent entry ten years before this, around the time when the first pointed-toe shoes (twelfth century onwards) were making their comeback as **winklepickers**.

UNDERWEAR

The first underwear for men and women, dating back to the invention of textiles (see p.35) or even earlier – think underpants made of bark or leaves – was the **loincloth** (also known as the *kaupinam*, breechcloth, *fundoshi*, etc.). **Pants-style underwear**, known as '*braies*' or '*braccae*' in Latin, were worn by tribes outside the Roman Empire from the first millennium BC onwards. Loincloths and underpants remained the standard underwear worldwide until the nineteenth century. Sporty Roman women wore bikini-like **breastcloths** ('*strophiae*'), with modern **bras** (featuring a pair of 'breast bags') making a surprisingly early appearance in fifteenth-century Austria. These predate by hundreds of years the creations of the late nineteenth century French corset-maker Herminie Cadolle and the US **bra patent** (1914).

CORSETS, SOCKS AND STOCKINGS

Though there is evidence of **corset** wearing in ancient Crete at least 3,500 years ago, the modern corset is said to date from the period when Catherine de Medici (Queen of France 1547–59) banned women with thick waists from attending court! Elasticated **girdles** replaced corsets in the mid-twentieth century. For centuries there was little distinction between **socks** and **stockings**. The first socks were Greek and made from animal hair (*c.* 750 BC). By Anglo-Saxon times (sixth century onwards) woollen socks had become longer, requiring **garters** to hold them up. **Hose** (the forerunner of stockings and pantyhose) were first made in the early medieval period. **Suspender belts** (garter belts in the USA) date from the nineteenth century. **Nylon stockings** ('nylons') first went on sale in Wilmington, Delaware, USA, on 24 October 1939, and Allen Gant (USA) produced the first **pantyhose** ('tights') in 1959.

HATS

The oldest **hat** may well be the 5,500-year-old copper **crown** discovered in a cave near the Dead Sea. A Russian-style bearskin **cap** (believed to have been made *c.* 3300 BC) was found with a frozen body in the Alps. The first reference to **modesty head covering for women** is in an Assyrian text from the thirteenth century BC (Iraq/Syria), when the practice of hiding the hair was reserved for high-born ladies, and those of lower social standing were punished for emulating the style of their superiors. Turkey issued one of the first modern **bans of the headscarf** in 1980. The earliest **fascinators** – the antithesis of the headscarf – originated in late sixteenth-century Europe and were revived in the 1990s. **Cloth caps**, or 'bonnets', appeared in fourteenth-century Britain, the military **tricorne** in seventeenth-century Flanders, the **top hat** in France in the late eighteenth century, and the bowler, a hard hat with a rounded crown, in 1849 (made by Thomas and William Bowler, UK). The robber's trademark headdress, the **balaclava**, began as warm clothing sent to British soldiers serving in the Crimean War (1854–6), though the phrase 'balaclava helmet' was not used until 1881. Finally, the **Stetson** – the 'Boss of the Plains' cowboy hat – was designed and manufactured in 1865 by ... American John Stetson, of course!

KEEPING WARM AND DRY

Protection against the weather begins with the first clothing (see p.34). More specifically, during the Iron Age (first millennium BC) Northern Europeans were making **woollen overgarments**. From 1587, they were known as '**jerseys**' because of their popularity with the fishermen of the Channel Islands. The word 'jumper' was not used for a garment until the mid-nineteenth century (UK). The Americans were calling jerseys '**sweaters**' by the end of the same century, and '**pullover**' entered the vocabulary in the 1920s (UK). The first **cardigan** (knitted waistcoat) was supposedly worn by the British Earl of Cardigan, leader of the calamitous Charge of the Light Brigade in 1854. Waterproof **oilskins** (including **sou'wester hats**) date from seventeenth-century Europe. **Capes**, probably medieval in origin, went out of fashion when Scotsman Charles Macintosh invented the rubberized **raincoat** (the 'mac', 1824).

A gentleman's Macintosh, from an 1893 catalogue

CHASTITY BELTS

The earliest reference to a chastity belt, with illustration, is to be found in *Bellifortis*, a fifteenth-century book on military engineering by the German writer Konrad Kyeser. Scholars now believe the reference was a joke. They also reckon the array of fanciful chastity belts on display in museums around the world are all fakes made in the late nineteenth and twentieth centuries. So the first chastity belt? It existed only in the minds of the prurient or obsessively virginal!

BATTLE DRESS

According to definition, the first **military uniform** was either the red cloak of the Spartan hoplite (*c.* 500 BC), the get-up of Roman legionaries from about AD 50, or the regimental dress of the French army after *c.* 1550. As with armies, the first **naval uniform** is open to question; it may be the blue-grey worn by sailors of Imperial Rome (first century AD) or the semi-standardized dress required of British naval officers from 1748 onwards, or (most reliably) the Royal Navy's 1857 order that standardized clothing throughout the service. The **khaki** (Urdu for 'dust-coloured') uniform appeared in the nineteenth century when members of the British army in India discoloured their white clothes with a mix of curry powder, coffee and mud to make it less conspicuous.

ARMOUR

The first documented **armour** (eleventh century BC) are the layers of rhinoceros hide worn by Chinese warriors. **Chain mail** originated in Etruscan Italy in the fourth century BC,

Chinese rhinoceros hide armour, 1852

and **plate armour** was worn by the ancient Greeks (first millennium BC). The earliest known request for a **bulletproof vest** was made in 1538, and the first protective garments of proved efficacy were the cuirasses worn by cavalry of the English parliament's New Model Army (1645), known as 'Ironsides'. The original **lightweight bulletproof vest**, capable of protecting the wearer against modern firearms, was the American Body Armor K-15 (1975 – see also p.37).

SMART AND FORMAL

Academic dress goes back to a ruling by Cardinal Stephen Langton, Archbishop of Canterbury (UK), in 1222, and Christ's Hospital School (UK) was the first to introduce a uniform (1553). The **three-piece suit** can be traced to the court dress worn at the time of King Louis XIV of France (1643–1715), which in turn echoed styles popular in Holland the previous century. **Waistcoats** were Persian and came to Europe with emissaries to the court of King Charles II (1660–85, UK), and the first person to leave the bottom button undone (to accommodate his spreading stomach) was Britain's King Edward VII (1901–10). **Morning dress** (now reserved mainly for formal weddings) was created in nineteenth-century Britain, while **frock coats** (now obsolete) were worn by German and Austrian officers of the armies fighting the French emperor Napoleon I in the early nineteenth century. In 1860, Edward Prince of Wales (UK) ordered a **dinner jacket** suit as an alternative to the more cumbersome tailed coats hitherto worn by gentlemen. **Neckties** appeared at the court of King Louis XIII of France (1610–43), who copied the idea from the lengths of cloth worn round the necks of his Croatian mercenaries. The first **women's suits** were European riding attire of the mid-seventeenth century, while **trouser suits** made their debut in Europe in the 1960s.

TAKING IT EASY

Shorts have their origin in the breeches and culottes of ancient times; it is suggested that modern-day shorts were first worn by the Gurkhas of the Nepalese army in the 1880s. The ubiquitous **jeans** were invented by Jacob Davis and Levi Strauss (1873, USA). There is a lineage from the waistcoat (see p.44), via the vest as an undergarment (early nineteenth century), to the **T-shirts** issued to the US Navy in the early twentieth century. The first recorded appearance of a **printed T-shirt** is in the 1939 film *The Wizard of Oz*, in which workers wear green T-shirts bearing the word 'Oz' (USA). **Miniskirts** can be seen in the artwork of ancient Egypt (*c.* 1390–1370 BC) and on dancers throughout the ages, but the modern minis turned heads on the streets of London (UK) and Paris (France) in the early 1960s. **Hoodies** have a similarly vague genealogy. Hoods (the word comes from the Anglo-Saxon for 'head') and cowls for monks were prevalent in medieval Europe, but we don't get the hooded sweatshirt until 1934 (USA), and the garment was not called a 'hoodie' (a word previously reserved for a type of crow) until 1991.

SPORTSWEAR

The first recorded sporting dress was undress – competitors at the original Greek Olympics (from 776 BC) competed naked! The only permitted male coverings were leather bandages on the hands of boxers, first seen in a Cretan fresco (*c.* 1500 BC). Female Greek athletes wore a simple short dress. Before the nineteenth century, athletic men and women generally took their clothes off to swim (thus swimming was considered a morally dubious activity) or, for sporting activities on land, they wore versions of everyday clothes. (The Roman 'bikini' – see p.40 – and the 1525 football boots of King Henry VIII of England are notable exceptions.) From the late eighteenth century, as the benefits of exercise were better understood and new sports invented, more sport-specific clothing and protective gear came along.

FASTENINGS

Shoelaces are as old as shoes themselves (see p.38) and some form of sewing and lacing as old as clothing (see p.34). **Buttons** were first sewn on around 2700 BC (India), and bronze **buckles** a few centuries later. Pinned **brooches** (widespread) appeared about the same time. For some reason **toggles** were employed in cold climates (*c.* 600 AD, Scandinavia, Russia and Canada), while the **hook-and-eye** is attributed to the more temperate British Isles of the fourteenth century. A form of snap fastener is found on the harness of the figures of the Chinese Terracotta Army (210 BC); German Heribert Bauer patented the modern version in 1885. **Cufflinks** (Europe) were first used in the seventeenth century, and **shirt studs** were needed with the arrival of the detachable shirt collar (1827, USA). The first **zip fastener** (zipper) is controversial, but our prize goes to Gideon Sundback (1913, Swedish-American). There's no such problem with **Velcro**, the invention of Swiss engineer George de Mestral, which was patented in 1955.

DOMESTIC TOOLS, FURNITURE AND GADGETS

CUTLERY AND CHOPSTICKS

Primitive **cutlery** – a lump of sharp flint or similar material – was used towards the end of the Stone Age (up to 2.4 million years ago), with **metal knives** (copper) being introduced around 4500 BC. **Table knives** (blunt-ended and for eating only) appeared in France around 1640. The earliest **forks** were made of bone (*c.* 2200 BC, China), though for years these implements served only for cooking and serving. Well-off households of the fourth-century Eastern Roman Empire were probably the first to have **table forks** at mealtimes. **Spoons**, made of bone, were first used for religious practices (*c.* 1500 BC, China). Not long afterwards, the Chinese fashioned bronze spoons as cutlery. The same culture gave us **chopsticks** as kitchen utensils (1766–1122 BC, Shang Dynasty) and eating tools (*c.* AD 220). The first **silver spoons** (the sort the fortunate have in their mouths at birth) were probably made in ancient Greece (fourth century BC or earlier).

TABLEWARE

Basic **pots** and **bowls** are among the earliest ceramics (*c.* 20000 BC, China, see p.17). The first **plates** – sturdy leaves or wooden trenchers or platters – were also in use many thousands of years ago (no one invented the leaf), and the first **ceramic plate** was made at least 3,000 years ago in China. In 2017, a Chinese **bronze plate** from approximately the same era – probably one of the first – fetched a record price at auction. People were drinking out of pottery **beakers** for millennia before Europe's

An ornate silver spoon, *c.* fourth century BC

Bell Beaker culture (*c.* 2900–1800 BC), and leather drinking vessels were first made around this time. The earliest pottery **mug** (beaker with a handle) is likely to have been made in the fifth millennium BC (Greece). Sophisticated Chinese created the **teacup** (*c.* 210 BC), though it remained for the English to give it a handle (1750, Robert Adam). Teacups have been resting on **saucers** since 1700. The Egyptians and Mesopotamians made **glass vessels** (and invented **cut glass**) around 1500 BC, a process made much easier with the arrival in the first century BC of the **blowpipe** for glass manufacture (Syria). The first **drinking glass** is attributed to the reign of Egypt's King Thutmose III (1479–1425 BC). **Paper plates** were made in 1867 (Germany) and **paper cups** in 1908 (USA). America was also responsible for the curse of the **plastic cup** (patented 1964).

LIQUID STORAGE

The Mesopotamians were making bronze containers that looked like **kettles** about 5,500 years ago. Cast-iron kettles followed Abraham Darby's introduction of sand casting in 1707 (UK); the first kettle **whistle** screeched in the USA in 1890, and the **electric kettle** first helped the British make tea in 1892. Water carriers in the form of pitchers, ewers and jugs go back at least 12,000 years. **Toby jugs**, supposedly named after the celebrated Yorkshire (UK) beer drinker Toby Philpot (Fill-pot!), appeared in the middle of the eighteenth century. Romans and Celts are each credited with making the first **barrels** approximately 2,000 years ago. Almost two millennia later (1934), Flower's India Pale Ale was shipped from the UK to India in the first **steel kegs**. Stainless-steel kegs were available in the 1950s and aluminium alloy ones in the 1960s. Early **bottles** were the leather skins of prehistoric times. The earliest **glass bottles** were made in South-East Asia around 100 BC and **plastic bottles**, the other infuriating environmental polluter, in 1947 (USA – see p.155). The **thermos** (vacuum) **flask** was invented in 1892 by Sir James Dewar (Cambridge, UK) but manufactured in Germany (1904).

FURNITURE

Since the beginning of civilization, people have used natural objects as furniture (tree stumps, rocks etc.). Seventy-seven thousand years ago, the inhabitants of KwaZulu-Natal, South Africa, may have made the first **bed** out of layers of plant material, but this hardly qualifies as furniture. The oldest extant furniture – stone **cupboards**, **shelves**, **seats** and beds – is found at Skara Brae (*c.* 3000 BC, Orkney, Scotland). The earliest evidence of **wooden furniture** (chairs, stools, storage boxes and raised beds) comes from ancient Egypt about 3100 BC. Though the wealthy and important had been carried around by slaves for centuries, the first **sedan chair** was hoisted onto the shoulders of its bearers in France or the UK in the second half of the sixteenth century. The first **deckchair** (steamer chair) is probably that patented by John Cham (1855, USA).

BED WARMING

The quest for a warm bed on a chilly night produced some interesting initiatives. The first and most obvious was to find a willing (and warm) companion. Failing that, in medieval Europe servants placed heated stones in their employer's bed and removed them before occupation. The warming pans – like frying pans with lids, filled with coals or hot water – were in use in mid-sixteenth-century Europe. (Hence the 1688 'Warming Pan Baby' story about Britain's fifty-four-year-old King James II and his second wife, twenty-nine-year-old Queen Mary: after a string of children died in infancy, the couple were no longer believed capable of producing a healthy child – and when they 'miraculously' did so in June 1688, the king's scandalized enemies said the baby was an imposter, smuggled into the queen's bed in a warming pan.) The rubber hot-water bottle (1875, UK) and the automatic electric blanket (1936, USA) were more efficient – and scandal-free.

BEDS AND CHESTS

Inventive Persians of *c.* 3600 BC made the first **waterbed** of goat skins filled with water; the first **airbed** (Lilo) was made in Massachusetts, USA, in 1889. **Camp beds** were used in classical times; the **futon** was crafted in eighteenth-century Japan, and the **sofa bed** patented in 1899 (USA). The Irish of the twelfth century are credited with the **four-poster bed**. Heinrich Westphal (Germany) devised the **innerspring mattress** in 1871; despite sleeping in comfort, he died in poverty after failing to profit from his invention. The **wardrobe** (to hang clothes rather than storing them in a box) appears to date from twelfth-century Europe, and the **chest of drawers** a few hundred years afterwards. **Couches** first appeared in ancient Egypt, morphing later into the **chaise longue** (sixteenth century, France) and the **sofa** (from *suffah*, the Arabic for a couch) in the 1620s.

MIRROR, MIRROR ON THE WALL

Even before the drop-dead gorgeous Narcissus fell in love with himself after seeing his reflection in a pool, we have wanted to see what we look like. Polished stone mirrors were made in Anatolia (Turkey) around 8,000 years ago. Metal mirrors came next, *c.* 4000 BC. The Roman author Pliny the Elder refers to glass mirrors in the first century AD, and by the end of the second century Romans were putting shiny metal behind the glass to give an even more accurate reflection of themselves. Fast-forward about 1,500 years to meet Justus von Liebig, one of Germany's most celebrated scientists. As well as giving us Oxo cubes and Marmite, in 1835 he learned how to deposit silver onto glass – thus creating the object capable of making Narcissi of us all: the modern silver-backed mirror.

GADGETS

The claw-type **can opener** (replacing a hammer and chisel!) was designed by Robert Yeates (1855, UK) – more than half a century after the introduction of canned food (see p.16). The **rotating wheel** opener was invented in the US in 1870. A **salad spinner** was patented in 1971 (France), though similar machines had apparently been around since the previous century. Other circular motion machines include the **juicer** (1936, Dr Norman Walker, USA) and **food mixer** (1856, hand-operated whisk; 1885, handheld electric mixer, Rufus Eastman; and 1908, electric standing mixer, Herbert Johnson – all USA). Frenchman Pierre Verdon designed the **food processor** – Le Magi-Mix! – in 1971.

MAKING TEA

The automatic **tea-maker**, patented in 1892 by Samuel Rowbottom (Derby, UK), involved a potentially lethal combination of clockwork alarm clock, gas ring and pilot light – hardly ideal for the bedside table. The first electric tea-maker, the **Teesmade**, was manufactured in 1933 (UK) and the more successful **Teasmade** went on sale in 1936. The first **teapots** were Chinese, perhaps from the Sung Dynasty (960–1279), and the contents were drunk through the spout. English **tea sets** were produced in the early eighteenth century (Queen Anne had a silver one), though the Chinese had been making sets with fewer pieces (i.e. no receptacles for milk or sugar) long before this.

A very British invention – a Goblin Teasmade, *c.* 1935

THE COFFEE CHALLENGE

Since time immemorial, **coffee** was made by simply mixing ground coffee beans with boiling water. Then, from France in 1710, came the first **infusion** – boiling coffee in a little fabric bag. Next, around 1780, came France's **coffee filter** and **percolator**. If versions of the latter were high-tech, they were nothing compared with the machine exhibited in Turin (Italy) by Angelo Moriondo in 1884: the **espresso machine**. Back in the kitchen, the call went out for something simpler and cheaper. Several Italians responded: Attilio Calimani and Giulio Moneta designed the **coffee press** (plunger) in 1929, and four years later Luigi De Ponti created the stove-top **moka pot** (1933). With the **electric drip coffee**

A 1965 advert for the stove-top Moka pot, invented and made in Italy

maker (1954, Germany) and Nestlé's **coffee pod** machine (1976, Switzerland), it seemed the stream of coffee-making inventions had run dry. But I wouldn't bet on it …

KEEPING CLEAN

SWEEPING

Brooms are as old as civilization (see p.7) and the earliest mention of them as witches' aviation tools was in 1453. Flat brooms were an American invention of the early nineteenth century and were used in tandem with another American invention, the **dustpan**, from 1858. Manual carpet sweepers began with Daniel Hess's **carpet sweeper** of 1860 (USA). The first **powered sweepers** arrived at your house on a horse-drawn vehicle. An 1898 version, driven by a petrol engine, *blew* the dust into a bag; the following year an electric motor was fitted for the first time. The year 1901 saw a significant improvement – the **vacuum cleaner** (UK and USA) – though it was still horse-drawn. The first household vacuum cleaner was made in Birmingham (UK, 1905); before long, it lost its market share to the extremely successful **Hoover** Suction Sweeper Company (1915, USA). The first **bagless vacuum cleaner** (Dyson, UK) went on sale in 1993.

WASHING

All early **washing machines** were hand-powered. These included an 'engine' of 1691, various eighteenth-century washing tubs and mills (all UK), and Nathaniel Briggs's 'Box Mangler' of 1797 (USA). A major design first came in 1851 when American James King patented the **drum washing machine**. A British washing machine-cum-mangle (for wringing out wet clothes) was patented in 1862. Between the two, in 1853, came the spring-loaded **clothes peg** (USA). Fifty-seven years later, the great breakthrough everyone had been waiting for appeared: the mighty Thor – the first **electric-powered** machine (USA). The rest, as they say, is history, as fully automatic machines and **washer-dryers** (1953, USA) followed. *Dreft*, the first **detergent**, was marketed in 1933 (Proctor and Gamble, USA), and the first **laundromat** opened in Fort Worth, Texas, USA, in 1934. A certain Monsieur Pochon (France) had designed a hand-driven **clothes dryer** in 1800, but the electric version of 1938 (USA) proved vastly more effective.

BATH TIME

Whether for ritualistic reasons or personal hygiene, humans first bathed wherever there was sufficient water, and by 3000 BC many early civilizations had built small but rarely private **bathrooms**. The Babylonians made the first recorded **soap** around 2800 BC, and the first large **public bath** (roughly 12m x 7m, with a depth of 2.4m) was constructed in the middle years of the same millennium at Mohenjo-daro (Pakistan). The earliest personal **bathtub** (*c.* 1500 BC) was probably that in the great palace of Knossos, Crete. **Metal bathtubs** (replacing wooden ones) date from the eighteenth century, and the cast-iron **enamel bath** (*c.* 1885) was the work of Scottish-American David Buick, who went on to pioneer motor cars. The Romans appear to have made the first **taps** (*c.* 100 BC); **mixer taps** were invented by the Canadian Thomas Campbell (1880). **Liquid soap** was manufactured in 1865 (USA), and Radox marketed **shower gel** in the 1970s (UK). Although better-off citizens in the ancient civilizations enjoyed a form of **shower** when slaves tipped buckets of water over them (*c.* 1000 BC onwards), and as a hand-pump shower patented in the UK in 1767 did not catch on, the modern shower was really born with the 'English Regency' model of *c.* 1810.

TOILET TALK

Mohenjo-daro (see above) also featured some of the first **toilets** as (perhaps surprisingly) did the remote Orkney settlement of Skara Brae, both dating from the third millennium BC. The ancient Greeks are the first people known to have used **chamber pots** (fourth century BC or earlier); Sir John Harington (*not* Thomas Crapper) invented the **flushing toilet** in 1596 (UK); ancient Rome had the first **public toilets**; and the **vacant/engaged bolt** was patented in 1883 (UK). The puritanical euphemism '**rest room**' dates from 1897. Though the **bidet** is supposed to have been a

French invention (the word means 'pony', something one sits astride!), it is first mentioned in Italy in 1726. In 1975, the Italians went a step further by making bidets mandatory. Toto, a Japanese manufacturer, marketed the **paperless toilet** – a sort of washer-dryer – in 1980. The first **railway carriage with a flushing toilet** (two, actually) was Pullman's 'Old No. 9' sleeper of 1859; an **airborne toilet** was installed in the pre-World War I Sikorsky Ilya Muromets (January 1914, Russia; see p.127).

NAPPIES AND PAPER

The first **nappies** (diapers) were probably made of leaves or hay. After that, any old bit of sacking or cloth sufficed. A baby's life became more comfortable with the **safety pin** (1849, USA) and the carer's life easier with the **disposable paper nappy** (1942, Sweden). Anal cleansing was initially performed, if at all, with the hand, sand, sponge, wool, leaves, small stones (the Prophet Muhammad recommended three) or by dipping into water. The Chinese, who had invented **paper** (see p.10), were the first to use it for toilet purposes (*c.* sixth century BC). Mass-market **toilet paper** dates from the middle of the nineteenth century (USA and elsewhere), and Andrex **wet wipes** (UK) from the mid-1990s.

RECIPES, COOKBOOKS AND MANUALS

The earliest written **recipes** are on Mesopotamian tablets of about 1700 BC. *De re coquinaria* (first century AD, Roman) is often said to be the first **cookbook**, though it was not printed until 1483. Arabic cookbooks date from the tenth century and Chinese from around 1330. One of the oldest **household management** books is *Le Ménagier de Paris* (*'The Householder of Paris'*, 1393), predating the better-known *Mrs Beeton's Book of Household Management* (1857–61, UK) by almost 500 years. The intensely Francophile *Larousse Gastronomique* (1938) was the first **encyclopedia of gastronomy**. Betty Crocker's ***Picture Cookbook*** (1950, USA) was the first to rely heavily on colour photography.

EATING OUT

The first well-documented places where passers-by could drop in for a drink and a meal were the **taverna** of ancient Athens which were flourishing around 400 BC. Smart **restaurants** (*bouillons*), where one went to get something out of the ordinary, began to open in Paris, France, around 1765. The **bistro**, also French, dates from 1884. Lyons opened their first distinctive art deco **Corner House** in 1909 (London, UK), and the Americans gave us the **fast-food restaurant** in 1921 (see p.22). The **pub** – an ordinary house where the public might drink ale (i.e. a public house) – appeared in early Anglo-Saxon Britain (fifth to sixth century BC), though the Angles and Saxons may have brought the idea with them from continental Europe. David Eyre and Mike Belben turned The Eagle (Clerkenwell, London, UK) into the first **gastropub** in 1991. **Bars** are a question of etymology – taverns were first described as 'bars' (i.e. barriers or counters) at the end of the sixteenth century.

PART III: HEALTH AND MEDICINE

EARLY DAYS

EXTRAORDINARY EGYPTIANS

Just as great apes self-medicate by eating certain plants, we can assume early members of the genus *Homo* did the same. What they consumed was, in its crude way, the first **medicine**. However, as the treatment (such as the Indian Ayurveda system, *c.* 4000 BC) mixed hit-and-miss herbalism with magic and spiritualism, it is probably more accurate to describe it as the first **alternative medicine**. For the first written **medical texts** we need to turn to the clay tablets of ancient Sumer (*c.* 1900 BC), and several papyrus documents from ancient Egypt (1800 BC onwards). Indeed, the innovative medical work of the ancient Egyptians was truly remarkable. They gave us the first **named physician** (Hesy-Ra), the first **female physician** (Merit-Ptah), **specialists**, **medical centres** ('Houses of Life', *c.* 2200 BC), **gynaecology** and **midwifery**, the first mention of **tumours**, the **brain**, the **pulse**, and **scientifically effective medicines** such as **castor oil** for constipation and **cannabis** for pain relief. It is thought, too, that they pioneered **cataract surgery** (using the 'couching' method).

ACUPUNCTURE

The origins of sticking needles into the body to cure illness or deaden pain are obscure. Many believe the practice began around 600 BC in China with the physician Bian Que, though some trace the practice back to the Shang dynasty (1766 BC onwards). Critics of these early dates are sceptical because the lack of steel would have meant piercing the flesh with needles of soft metal such as copper, gold or silver – or even with thorns, shards of flint or pointed slivers of bamboo. Then there are those who suggest that the positions of the sixty-one tattoos on the body of Ötzi the Iceman, a human preserved in Tyrolean ice for over 5,000 years, indicate a primitive form of acupuncture. Be that as it may, the practice spread beyond China to Korea in the sixth century BC, and first came to Europe – and thence to America – in the seventeenth century.

GREECE

In medicine, as in so many things, the ancient Greeks came up with a number of remarkable firsts. The outstanding figure – sometimes called the first **modern physician** – was Hippocrates of Kos (*c.* 460 to *c.* 370 BC) who attempted to separate illness from unscientific mumbo jumbo. He is also said to have introduced the concept of **lifestyle medicine** ('walking is the best treatment') and **medical ethics** (the modern Hippocratic Oath is a tribute). Herophilos (335–280 BC), who distinguished between **arteries** and **veins**, is known as the first **anatomist**; he and Galen (AD 129 to *c.* 200) are credited with being the first to realize that the **brain** was both the centre of intelligence and the hub of the nervous system. Herophilos was a pioneer of **dissection** – though in modern eyes his reported **vivisection** of live criminals detracts somewhat from his reputation. Mention, too, must be made of Erasistratus (*c.* 304 to *c.* 250 BC), another member of the Alexandria anatomy school and alleged vivisectionist, who realized the **heart** is a pump.

ISLAM

The torch lit in ancient Egypt and the classical world was kept alight in the Middle Ages by Arab scholars who fused European medical knowledge with that of Asia to produce several remarkably prescient texts. The Persian polymath Abū Bakr Muhammad ibn Zakariyyā al-Rāzī (AD 854–925), hailed by some as the 'Father of **Paediatrics**, **Psychology and Psychotherapy**' (see p.78), was the first to distinguish **measles** from **smallpox**. He also described **cataract removal**, a practice that may have begun in ancient times. Ibn al-Nafis (1213–88) discovered **pulmonary circulation** centuries before the more complete and detailed description published in 1628 by the Englishman William Harvey (1578–1657). Basra-born Hasan Ibn al-Haytham (*c.* 965 to *c.* 1040) was the first scientist to explain the **eye** as an instrument and **vision** being the result of light passing from an object into the eye.

SCIENTIFIC MEDICINE

Three firsts mark the beginnings of modern medicine.

1) The publication in 1543 of *De Humani Corporis Fabrica* ('On the Fabric of the Human Body') by the 'Father of **Modern Anatomy**', the Fleming Andreas Vesalius (1514–64). He gave us the first illustration of the **skeleton** as the body's framework, and of the position and function of **muscles**.
2) The first clear statement, by the British statesman Francis Bacon (1561–1626), of what came to be known as the empirical '**scientific method**' – i.e. knowledge and theory should be based not upon given truths but upon tangibly verifiable and constantly reassessed facts.
3) Using microscopes (see p.158) of his own design and making, Dutchman Antonie van Leeuwenhoek (1632–1723) revealed the hidden world of **bacteria**, **spermatozoa**, **red blood cells** and other microscopic organisms.

It is no exaggeration that on these three pillars rest all the remaining firsts in this section.

DRUGS

FIGHTING PAIN

Poppy seeds and **alcohol** were first prescribed to alleviate pain in ancient times, but the first recommendation that opiates, in the form of **laudanum**, be taken for analgesic purposes was made by the Swiss physician Paracelsus (1493–1541) in about 1525. **Morphine** was isolated by the German chemist

Friedrich Sertürner around 1804 and went on sale three years afterwards. The Chinese surgeon Hua Tuo (*c.* AD 140–208) is credited with the first recorded use of **cannabis** as an anaesthetic, though the Egyptians almost certainly employed it before this (see p.59). **Nonsteroidal anti-inflammatory drugs** (NSAIDs) began with **aspirin**, isolated as **salicylic acid** (the active ingredient of willow bark, recommended by Hippocrates 2,000 years earlier!) by Charles Gerhardt (1852, France). The German company Bayer marketed it in 1899, and it was mixed with sodium bicarbonate and citric acid to make **Alka-Seltzer** (1931, USA). **Ibuprofen** became available in the 1960s and **naproxen** a decade later. Over half a century passed between the launch of a toxic form of **paracetamol** (Antifebrin) in 1886 and its sale as a relatively harmless drug in 1950.

GOING UNDER

The word 'anaesthesia' was coined by the American writer Oliver Wendell Holmes in 1846. However, as previously noted (see opposite page), **opium**, alcohol, cannabis, hemlock and other herbs (individually or mixed) had long been taken as analgesics and sedatives. Large doses were frequently fatal. The opium poppy, it is suggested, was cultivated by the Sumerians as long ago as 3400 BC. The first reasonably reliable record of **general anaesthesia** comes from second-century China, where Hua Tuo (see above) reportedly knocked out patients before surgery by administering a mysterious potion known as *mafeisan* (possibly a mix of herbs, wine and cannabis). Whether any of them recovered is not recorded. Hanaoka Seishū of Osaka, Japan, combined study of Hua Tuo's compound with a knowledge of Western science to

COCAINE

Realizing early on that life was, in the words of the English philosopher Thomas Hobbes, 'nasty, brutish and short', *Homo sapiens* sought ways of alleviating their misery by taking substances that eased their grip on reality. The preferred option of South Americans was chewing on the leaves of the coca plant. Though this habit is said to be thousands of years old, the first documented use comes from the mummified corpses of Inca children who, in the 1400s, were made high before being sacrificed to the gods. In the following century, Spanish invaders not only partook of the drug but made money by taxing it. Its first advocacy for medicinal purposes – specifically for healing rotten wounds – came shortly afterwards. In 1855, the German chemist Friedrich Gaedcke isolated the effective alkaloid. Another German, Albert Niemann, came up with the name 'cocaine' (from the coca plant) when working on the alkaloid for his PhD (1860). Before long, the substance was everywhere: in cigarettes, in the original recipe for Coca-Cola (see p.31) and even, it is alleged, in a flask carried by Pope Leo XIII. Owing to its highly addictive properties, the drug received its first international ban at the Paris Convention for Limiting the Manufacture and Regulating the Distribution of Narcotic Drugs (1931).

produce *tsūsensan*, an anaesthetic powerful enough for him to carry out the first **reliably documented surgery** (a partial mastectomy) under general anaesthesia in 1804. Meanwhile, in the West, scientists and doctors were focusing on inhaled anaesthetics: **diethyl ether** (supposedly discovered in 1275, synthesized in 1540, and used as an inhaled anaesthetic in 1846, USA), **nitrous oxide** ('laughing gas', first made in 1772 and used for a surgical procedure – tooth extraction – in 1844, USA) and **chloroform** (used for general anaesthesia in 1847, UK). The first **intravenous anaesthetic** (sodium thiopental) was synthesized in 1934. German chemists manufactured **methadone**, the first man-made opiate, in 1937, and **meperidine** (aka Pethidine) in 1939.

THE SYRINGE

In the world of syringes, definition is important. The earliest reference to a squirting instrument with a squashy bulb and a narrow tube comes from the Roman author Aulus Cornelius Celsus (first century AD). In the modern era, several scientists considered the possibility of injections. One was Sir Christopher Wren (1632–1723, UK), who experimented on dogs using bladders and goose quills. But the **hollow metal needle syringe** – the type employed nowadays – was not manufactured until 1844, when the Irish doctor Francis Rynd injected a woman with morphine. Nine years later, the Scottish doctor Alexander Wood (and perhaps, at about the same time, the French doctor Charles Pravaz) came up with a **hypodermic syringe** with a hollow needle fine enough to pierce the skin. The **all-glass**, easily sterilized hypodermic syringe appeared first in 1946 (UK), a **plastic disposable** version in 1949 (Australia), and a syringe that separated from the needle after use (thereby reducing reuse and cross-contamination) in 1989 (Spain).

GERM WARFARE

A plague doctor in his mask

It took an astonishingly long time to discover the link between filth and infection. A few medics of the ancient and medieval worlds came close to an understanding when they suggested disease was carried by invisible 'seeds' (rather than '**bad air**'), but they lacked the scientific know-how to support the idea. What became known as the '**germ theory**' began with the work of the German Jesuit Athanasius Kircher on Roman plague victims in the middle of the seventeenth century; it was given a further boost by Antonie van Leeuwenhoek's discovery of **microbes** (see p.62). After Agostino Bassi, from the Italian city of Lodi, had proved that disease was carried by **microorganisms** (1813), the Hungarian obstetrician Ignaz Semmelweis (1818–65) insisted that every doctor in his department who dissected cadavers **washed with soap and chlorine** before attending women in childbirth. Though he was the first doctor to demonstrate the correctness of the germ theory and the concomitant practices it required, his work was largely ignored and he died in a mental asylum – of blood poisoning. Nevertheless, these developments led to the **pasteurization** process of the Frenchman Louis Pasteur (1822–95) and the work of Scotsman Joseph Lister, the 'Father of Modern Surgery', the first general surgeon to insist on **sterile conditions** in operating theatres (1870s).

DIABETES

It should come as no surprise to readers who have browsed thus far to learn that the earliest reference to **diabetes** comes from ancient Egypt. In 1552 BC, the physician Hesy-Ra (aka 'Great One of the Ivory Cutters' – a top dentist? See p.97) wrote that frequent **urination** was a symptom of a strange emaciating disease. The wee link was confirmed with the coining of the word 'diabetes', meaning syphon in classical Greek, *c.* 250 BC. In early modern times, the 'pissing evile' was diagnosed by European 'water tasters' able to identify diabetic urine by its sweet taste. Hence the addition of 'mellitus' (honey-tasting) in 1675 to give us the disease's full name, ***diabetes mellitus*** (DM). The role of the **pancreas in diabetes** was discovered by the German scientists Joseph von Mering and Oskar Minkowski in 1889. During World War I, the Romanian professor Nicolae C. Paulescu identified the hormone later called '**insulin**'. As a result of the work of a Canadian-Scottish team comprising Frederick Banting, Charles Best, James Collip and John Macleod, in 1922 the fourteen-year-old Canadian Leonard Thompson became the first diabetic to be **successfully treated with insulin**. **Synthetic insulin** was engineered in 1978 and went on sale in 1982 (both USA).

ST. MARTIN'S STOMACH

The Dutch scientist Jan Baptist van Helmont was the first to describe digestion as a chemical process (1648). The accuracy of this observation was not proved until 1822, when the US army surgeon William Beaumont began to treat a Canadian fur trader, Alexis St. Martin, who had been left with a gaping stomach wound after a shooting accident. Beaumont tied bits of food to lengths of string, inserted them into his patient's stomach and observed, when he pulled them out, how much they had been digested. St. Martin was understandably annoyed to be used as a piece of lab equipment and ran away to Canada. Beaumont had him brought back and continued his work, demonstrating how a cupful of acid taken from St Martin's stomach could 'digest' food outside the body. Beaumont, the 'Father of Gastric Physiology', published his findings in 1838. St. Martin returned to Canada, where he died in 1880 aged seventy-eight.

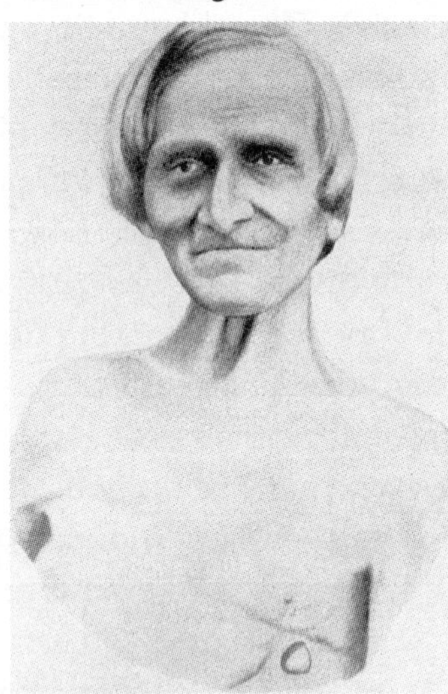

Alexis St. Martin, 1912

TUMMY TROUBLES

Though the physical properties of the **mouth**, **stomach and intestines** had been known for millennia, before the nineteenth century there was little understanding of their chemistry. The pioneering work of William Beaumont (see p.68) was crucial. Before that, the ancient Sumerians had prescribed the first **antacid** (*c.* 2300 BC, comprising an effective combination of milk, peppermint and **sodium carbonate**). **Magnesium hydroxide** was first employed as an antacid by the Irish physician James Murray in 1829. Forty-three years later, Englishman John Phillips marketed it – highly successfully – in the palatable form of **Milk of Magnesia**. In the next century, British scientists came up with **cimetidine** (1977), a drug that inhibited acid production before its painful effects were felt. All this time, the medical profession was blaming **stomach ulcers** on acid production (related to diet, stress etc.). Not until 1982 did the Australian physicians Robin Warren and Barry Marshall identify the bacterium that caused ulcers, though it took many years for the profession to accept their findings. Two other intestinal firsts are worthy of note. A successful **appendicectomy** was carried out in 1735 by the exiled French Protestant Claudius Amyand, working in St George's Hospital, London; and **peristalsis** was discovered in the 1890s by Ernest Starling (who also coined the word '**hormone**') and his brother-in-law William Bayliss of University College, London.

ANTIBIOTICS, THE 'WONDER DRUG'

The story of Alexander (later Sir Alexander) Fleming's accidental discovery of **penicillin** in 1928 (UK) is too well known to need retelling. It is far less known that, two millennia earlier, ancient civilizations in China, Egypt, Serbia and elsewhere had stumbled across the healing properties of **mouldy** (i.e. containing a primitive form of antibiotic) **bread**. Fleming's 'first' is also open to question. In 1870, Sir John Scott Burdon-Sanderson (UK) noted how mould inhibited bacteria growth, and in the 1890s, German scientists Rudolph Emmerich and Oscar Löw invented Pyocyanase, the first **antibiotic drug**. Moreover, Oxford scientists Howard Florey and Ernst Chain (UK) did not isolate penicillin, the effective drug in mould, until 1939. It was first used on a patient in 1941 (UK). The first **broad-spectrum antibiotic**, Aureomycin, was discovered in the USA (1945). The earliest warning against 'superbug' **antibiotic resistance** was made in 1954, and not until 2018 was an antibiotic developed that, it was claimed, could tackle known superbugs.

ANTIDEPRESSANTS

The treatment of **depression** over the years makes sorry reading. The first references to psychiatric disorder, said to stem from 'possession' by evil spirits, are found in the ancient literature of Babylon and China. The Greeks were the first to link mental health to physical health, but most adhered to the idea, effectively propagated by Galen (see p.61), that depression stemmed from an imbalance of 'humours' in the body. Robert Burton's *Anatomy of Melancholy* (1621) was an early effort at identifying the social causes of depression. The illness was first distinguished from **schizophrenia** in 1895 when Sigmund Freud (Austria) was pioneering **psychoanalysis**. In 1924, Louis Lewin (Germany) classified drugs and plants that had psychoactive properties; in the 1930s **convulsion therapies** began, using drugs as well as **electric shocks** (ECT). Modern drug treatment dates from the use of **lithium** (1949), with the widely prescribed fluoxetine (**Prozac**) available from 1986.

HALES'S HORSE

After leaving Cambridge, the brilliant Rev. Stephen Hales (1677–1761) conducted numerous important scientific experiments in his parish of Teddington, Middlesex, UK. One of the most significant involved inserting a brass pipe into an artery of a horse that was due to be put down. The pipe was attached to a nine-foot glass tube, fixed vertically. When a clip on the artery was released, blood pumped into the glass tube and rose to a height of eight feet plus a variable number of inches – the first measurement of blood pressure. The height of the blood fluctuated between two and four inches with each beat of the creature's heart. (Incidentally, using modern calibrations, the horse's systolic blood pressure was 185 – high for both a horse and a human.)

BLOOD PRESSURE

References to what was called '**hard pulse disease**' (i.e. hypertension that can be detected in the pulse beat) go back as far as the legendary Yellow Emperor of China of the middle years of the third millennium BC. The first treatment, advocated by Hippocrates (see p.60) and others, was simple: drain off some of the patient's blood by **opening a vein**, or allow **leeches** to do the job for you. This was pretty crude stuff, and blood pressure was not **measured** accurately until 1733 (see p.71). At the end of the century, the British physician William Withering (1741–99) discovered the beneficial effects of **digitalis** (foxglove) on heart and circulatory problems. The next major first was Frederick Akbar Mahomed's use of the new **sphygmograph** (invented in Germany, 1854, as the first non-intrusive pulse/blood pressure measuring device) to note that **hypertension** was not necessarily related to kidney disease. The Italian doctor Scipione Riva-Rocci devised the modern cuff-type blood pressure measuring device, the **sphygmomanometer**, in 1896. Another fifty years passed before the appearance of an acceptable oral **diuretic** (1957, chlorothiazide, UK/USA), and James Black's synthesis of propranolol and pronethalol (1964, **beta blockers**, UK). In 1959, doses of **hydrochlorothiazide** (a diuretic to lower hypertension) became available, and amlodipine besylate (**Norvasc**, which lowers blood pressure by increasing the size of the arteries) went on sale in 1990.

ANTI-INFLAMMATION

As in so many areas of medicinal drugs, the first use of anti-inflammatory substances – **myrtle** and **willow bark**, both containing **salicylic acid** (see p.63) – dates from ancient times, with the earliest written evidence coming from Sumer (*c.* 2500 BC, Mesopotamia/Iraq). In the modern era, **aspirin** reigned supreme until the arrival of **nonsteroidal anti-inflammatory drugs** (NSAIDs, see p.63) in the 1950s. **Indometacin** (Indocid) appeared in 1965; **ibuprofen**, developed as a safer alternative to aspirin, was available by 1969, and **naproxen** in 1976.

STEROIDS

Nothing spreads faster than a good story, no matter whether it's true or not. A startling example is how athletes in ancient Greece ate – some versions say 'chewed on, raw' – the testosterone-rich testicles of rams or bulls to enhance their sporting performance. The whole thing is myth – fake news from the ancient world. Reliable demonstration of **chemical communication** between the organs of the body was first made by the German scientist Arnold Adolph Berthold in 1849. The British pair of Ernest Starling and William Bayliss discovered the first such chemical, **secretin**, in 1902, coining the word '**hormone**' three years later (see p.69). **Steroid hormones** were identified and isolated in the 1920s and 1930s. Surprisingly, **testosterone**, the primary male sex hormone, was both discovered and recreated in the laboratory (the first **anabolic steroid**) in the same year, 1935. Experiments involving the **injection of anabolic steroids** into human beings for therapeutic purposes (initially as a treatment for depression) started in 1937. The first documented case of **use by an athlete** comes from 1954 (Russian weightlifters). A **reliable test** for their presence was available in 1974, and they were **outlawed** by the International Olympic Committee in 1976.

PHARMACIES AND PRESCRIPTIONS

An old apothecary, 1651

At the time of the Babylonian king Hammurabi (1792–1750 BC), traders in medicinal drugs gathered in a certain street in the city of Sippar – the earliest reference to **pharmacy shops**. The first written **prescription**, we are told, is even older: it survives on a Mesopotamian clay tablet from *c.* 2100 BC. There is a **picture of a pharmacy** on the wall of an Egyptian tomb of *c.* 1400 BC. By the time of the Greek playwright Aristophanes (*c.* 446 to *c.* 386 BC) we have specific mention of the pharmacist: *pharmakopôlos*. **State regulation** of drug stores began in ninth-century Baghdad. In 1240, Holy Roman Emperor Frederick II decreed separation between the profession of **physician** and **apothecary**. Medieval Venice was the first state to demand that the ingredients of medicinal preparations be **made public**. British chemist John **Boot** opened his business in 1849, the humble beginnings of what would grow into the first **chain of pharmacies**.

TOBACCO

Tobacco was being **cultivated** in Mexico 3,400 years ago, though no doubt the wild plant was chewed or its smoke inhaled long before that. The earliest known tobacco **pipes**, over 5,000 years old, were uncovered in burial sites in the Mississippi Valley (USA), while the oldest **bongs** (water pipes or hubble-bubbles, supposedly used for smoking cannabis or opium) were found in Russia. Tobacco first left America in a Spanish ship in 1528. Soon afterwards, Bartolomé de las Casas (*c.* 1484–1566) described the product's pernicious **addictive** qualities. The Frenchman Jean Nicot (yes, the man who gave us *nicotine*) introduced the 'sacred herb' into France (1559), where it received rave reviews for its medicinal and relaxant qualities. Britain's King James VI and I was less approving, and his famous *Counterblaste to Tobacco* (1604) is the first written warning about the dangers of smoking.

Native Americans rolled the first **cigars** from dried tobacco leaves and the first **cigarettes** using reeds or plant leaves as wrappers. Europeans were employing **roll-your-own** paper wrappers by the end of the seventeenth century. When the Mexican Juan Nepomuceno Adorno invented a **cigarette-making machine** in 1848, the extraordinary global smoking phenomenon was well and truly launched. Where smoke led, profit and fashion swiftly followed. A German made the first **lighter** in 1823, an Englishman the first friction **match** in 1828, and a Swede improved on this sixteen years later with the **safety match**. Following the construction of a machine able to turn out thousands of cigarettes a day (1881, USA), the industrialized product was sold in **packs**. Cigarette **boxes** and **cases** became fashionable about the same time, though the word '**ashtray**' did not appear until 1926, a year after the first **filter** cigarette went on sale.

Dr Isaac Adler (USA) conducted pioneering research into the link between **lung cancer** and smoking as early as 1912 and, in a rare positive for the regime, German Nazi doctors supported Adler's conclusion. This evidence was largely ignored, and it was only when the British physiologist Richard Doll published a study (1948) linking smoking to **ill-health** that the medical community started to take the problem seriously.

A Russian **ban** on smoking (1634) is the first recorded, and Nazi Germany saw the first modern embargo on smoking in public places. The rest of the world was slow to follow. In 1965, the UK prohibited **cigarette advertising** on TV, and the US required a **health warning** on cigarette packs. Aspen, USA, was the first city to restrict smoking in **restaurants** (1985), and Ireland was the first country to introduce a total smoking ban in all **workplaces** (2004). In 2010, Bhutan became the first country to outlaw tobacco completely.

Once the harmful effects of smoking were clear, scientists set about devising products to help smokers quit. Their work focused on two areas: finding alternative ways of delivering addictive nicotine, and satisfying the pleasurable sensation of sucking. First came **nicotine chewing gum**, launched in 1971 (Sweden), and on sale in Switzerland in 1978, the UK in 1980, and the US in 1984. The product satisfied the craving for nicotine and partially met the desire for oral comfort.

Nicotine **patches** (available in 1991), **nasal sprays** (1994), **inhalers** (1996) and **under-the-tongue lozenges** (1999, all Sweden) also met the first criterion. In 1963, Herbert A. Gilbert (USA) invented the **e-cigarette**. His device met the desire to suck on something but its 'smoke' of flavoured steam failed to meet the need for nicotine – and the product failed. Forty years later, the Chinese Hon Lik (who graduated from a college of traditional Chinese medicine) made good the shortcoming with his nicotine-delivering e-cigarette or '**vape**' that was soon selling worldwide in many different forms.

MENTAL HEALTH

MADNESS

The practice by our ancestors 8,500 years ago of skull boring and drilling holes in the skull suggests that mental illness has always been a feature of the human condition. It is thought that boring (**trepanning**) – the first known treatment for psychiatric disorder – was carried out to release evil spirits disturbing the patient's brain. The earliest written evidence of mental illness comes from China *c.* 2700 BC; it explains how both physical and mental disorders appear to stem from an imbalance between the complementary forces of yin and yang. Amid all the mumbo-jumbo about 'madness', evil spirits and so on, two thinkers of the ancient world stand out. Hippocrates (see p.60) was the first to make a scientific **classification of mental disorders**, including paranoia and melancholia. Another Greek, Asclepiades (*c.* 124–40 BC), broke new ground by advocating that the insane be **treated humanely** with calm talk and massage. Aretaeus (first century AD) pointed the way towards **psychosomatic medicine** by linking a patient's physical and mental changes. The torch of enlightenment then passed to the Middle East, where Ahmad ibn Tulun built what may have been the world's first **mental hospital** (AD 872, Cairo). The Persian physician Abū Bakr Muhammad ibn Zakariyyā al-Rāzī (see p.61) was one of the first to see the **brain as the seat of mental illness**, and, working in Baghdad, he pioneered **psychotherapy** as director of a **psychiatric ward**. Another Persian, Abu Zayd al-Balkhi (AD 850–934), conducted an early form of **cognitive therapy**.

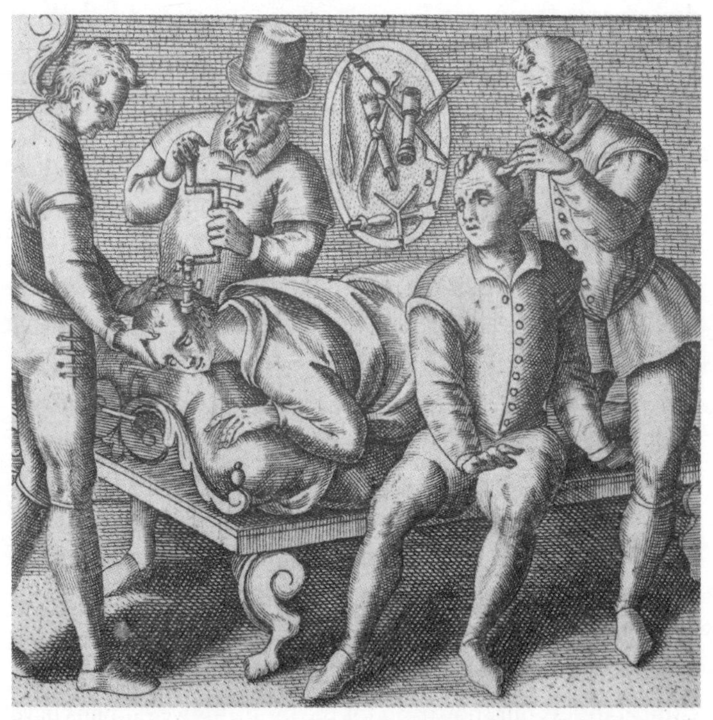

Trepanning – a crude and painful treatment for diseases of the mind

TOWARDS UNDERSTANDING

The practice of locking up the insane in **institutions** appears to date from the Middle Ages, with special cells in Paris's Hôtel-Dieu hospital (founded AD 651) among the first. Not until the Madhouses Act of 1774 were such institutions subject to any form of **regulation** in the UK. A major change in attitude – viewing those suffering from mental illness as **patients who might be cured** – began around 1790, notably with the work of the Frenchman Philippe Pinel (1745–1826), the 'Father of Modern Psychiatry' and the first to describe **schizophrenia** (1809). Other categories followed: **monomania** (1810), **dipsomania** (1829) and **kleptomania** (1830). The **psychology of sex** was first fully analysed in *Psychopathia Sexualis* (1886, Richard Freiherr von Krafft-Ebing, Germany – the Latin was a deliberate attempt to put off lay readers seeking titillation), which gave us terms such as **sadist**, **masochist**, **homosexuality** (then regarded as an illness), **necrophilia** and **anilingus**. Sigmund Freud (see p.71) introduced the concept of **psychoanalysis** (the word first appeared in print in 1896). The earliest written distinction between **psychoses** (severe disturbances) and **neuroses** (relatively mild conditions) was made *c.* 1913. The American Psychiatric Association's widely respected ***Diagnostic and Statistical Manual of Mental Disorders*** (DSM) was first published in 1952, and homosexuality was deleted from its list of mental illnesses in 1975. The term '**bipolar**' largely replaced 'manic depression' from 1980 onwards. The introduction of **psychotropic drugs** is dealt with above, leaving the last word to the Hungarian-American academic Thomas Szasz, one of the first to claim that what we call 'mental illness' is **not really a clinical illness** but shorthand for the problems we humans have with the experience of living (*The Myth of Mental Illness*, 1961).

SHOCK THERAPY

Other therapies having failed, or produced only transitory results, in the 1920s doctors experimented with shock treatments to break a patient's consciousness, thereby releasing them from a cycle of mental distress. Fever therapy, for treating paralysis brought on by tertiary syphilis, was first tried in 1919 by deliberately giving patients malaria. Insulin shock therapy began in 1927, when the Austrian-American psychiatrist Manfred Sakel zapped patients with a massive insulin dose to put them into a coma. Even more dramatic were electroconvulsive therapy (ECT), in which a current of up to 460 volts was passed through the patient's brain (1938), and lobotomy (cutting out the frontal lobes of the brain). This operation was first performed in 1935, and fourteen years later the pioneer Portuguese surgeon, Egas Moniz, was awarded a Nobel Prize for Medicine. The procedure is no longer carried out and there have been calls for Moniz's prize to be withdrawn.

SURGERY

BEFORE ANAESTHETICS

Human beings have been hacking into and cutting bits off each other since before the dawn of civilization. **Circumcision** – said to be the oldest surgical procedure – has been practised for over 15,000 years, possibly at first to mark the inferior status of a defeated enemy. A historical record of circumcision comes from an ancient Egyptian image of *c.* 2400 BC, though the god Ra is said to have self-circumcised long before that. Seven-thousand-year-old skeletons from French caves indicate that **trepanning** (see p.77) and **amputation** were first carried out in the Stone Age. The Sumerian Code of Hammurabi (1754 BC) mentions **tumour surgery**, and by AD 200 **mastectomies** and excisions of **breast cancers** were being performed in the Romano-Greek world. The first written references to **caesarean section**, one Chinese and the other Persian, date from *c.* 1000 BC; a more reliable account is that of the birth of the Indian emperor Bindusara *c.* 320 BC. Scholars are unsure whether the man who gave his name to the procedure, the Roman Julius Caesar, really was cut from his mother's womb. The first historical confirmation of **lithotomy** (surgical removal of stones from organs such as the bladder and kidneys) is in the work of the Greek physician Hippocrates (see p.60). Galen (see p.61), another Greek pioneer, advocated **catgut sutures** to bind up the wounds left by the primitive surgeons of his day. The earliest record of a successful **appendicectomy** is of that performed by the French surgeon Claudius Amyand in 1735 at St George's Hospital in London, UK.

SAFELY UNDER THE KNIFE

With the advent first of painless surgery (see p.63), then of antiseptic surgery (see p.67) in the mid-nineteenth century, surgeons were finally able to cure, where their predecessors had often killed. The year 1879 saw the first successful **removal of a brain tumour** (UK), 1880 the first **thyroidectomy** (Germany), and 1883 the first **salpingectomy** (removal of a fallopian tube, UK). Successful **cardiac surgery** was performed in 1895 (Germany) and **open-heart surgery** thirty years later (UK). Though records of **tonsillectomy** stretch back to the first century BC (when it was recommended, among other things, as a cure for bed-wetting), and numerous surgical devices had been invented to perform the operation, not until 1909 (USA) was it first regarded as a safe procedure. **Craniopagus twins** (conjoined at the head) were successfully separated for the first time in 1955 (USA), and a breakthrough **coronary bypass** operation was performed in 1967 (USA). The first successful **auto-appendicectomy** (surgically removing one's own appendix) was performed by the Russian Antarctic explorer Leonid Rogozov in May 1961. Despite understandable weakness and nausea, he had the rotten appendix out and the incision stitched up in just two hours.

'UNSEX ME HERE'

Long before Lady Macbeth asked the 'spirits that tend on mortal thoughts' to unsex her so that she could more easily commit murder, people had been conducting crude and usually cruel operations to prevent sexual intercourse and conception. Human **castration** in the ancient world of the

Middle East and China (where penis as well as testicles were removed) dates from at least the first millennium BC, with the earliest reference from Sumer in the twenty-first century BC. The first **medical description** of castration was written in the seventh century by the Byzantine physician Paul of Aegina. **FGM** (female genital mutilation or female infibulation) probably originated in north-east Africa in pharaonic times – the mummified remains of infibulated Egyptian females date from the fifth century BC. Female sterilization by **tubal ligation**, first reported in the USA in 1881, was rendered relatively safe and effective by a procedure introduced in 1930. (For **vasectomy**, see p.100.)

REPLACEMENTS AND TRANSPLANTS

The Chinese may have written about heart transplants around 300 BC, but – apart from peg legs and hooks for hands – modern transplant surgery dates from the second half of the twentieth century. Before that, the main developments were the first pair of **moveable prosthetic hands** (1508), fashioned from iron and worn by the German mercenary Gotz von Berlichingen, and the French army surgeon Ambroise Paré's construction of the first **articulated prosthetic leg** (*c.* 1536). Using ivory and nickel, in 1890 the German surgeon Themistocles Glück implanted the first **artificial knee**; the following year he followed up with the first **artificial hip**. Two years afterwards, the French surgeon Jules-Émile Péan fitted the first **artificial shoulder joint**. Improvements followed the availability of stronger, almost frictionless materials. Successful organ transplant surgery began with a **kidney transplant** between identical twins in 1954 (USA). Thirteen years later, techniques of overcoming rejection of

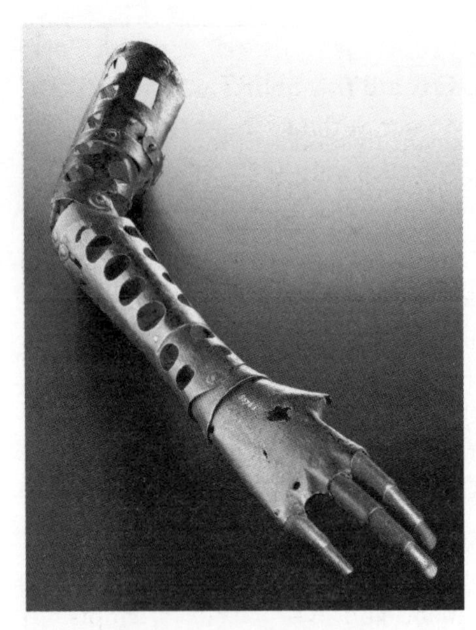

An articulated artificial right arm, 1501–1550

implanted tissue enabled the first **liver transplant** (USA). Most attention went to the first **heart transplant operation**, carried out by the South African surgeon Christiaan Barnard in 1967. His patient, who received the heart of a twenty-five-year-old woman, survived for a little over two weeks. The first **heart–lung transplant** took place the next year (USA). By the twenty-first century, such operations were becoming routine and surgeons were moving on to ever more complex procedures.

2010: the first **full face transplant** (Spain)

2011: the first successful **double leg transplant** (Spain); the first successful **womb transplant** (Turkey); the first **hand transplant** (UK); the first **double arm transplant** (USA)

2014: the first **penis transplant** (South Africa)

2015: the first **skull and scalp transplant** (USA)

UNDONE BY A SHIRT

The first female surgeon in modern times may well have been the Swiss physician Enriqueta Favez. After her undocumented birth in Switzerland in around 1791, at the age of fifteen Favez was married off by an uncle. She was widowed three years later. Dressing as a man, she then qualified in medicine at the Sorbonne, Paris, and served as a French army surgeon until captured by British troops in the Napoleonic Wars. On her release, she made her way to Cuba, where (still as a man) she married a woman and worked as a surgeon. A nosy servant revealed Favez's true identity when she found her mistress drunk – with her shirt undone! Imprisonment and two suicide attempts followed. Ever resourceful, the resilient Favez fled to New Orleans where she joined a community of nuns. At the time of her death in 1865 she was serving as their Mother Superior.

WOMEN IN MEDICINE

After the pioneering work of ancient Egyptian women (see p.59), it was virtually impossible for women anywhere to receive a serious, scientific medical education before the nineteenth century. Exceptions were the Greek physician Metrodora, the first woman to write a **medical treatise** (*c.* AD 200–400); the seventh-century medical worker Rufaida Al-Aslamia, the first female **Muslim nurse**; and the Italian intellectual Dorotea Bucca (1360–1436) of the University of Bologna, the first woman to **teach medicine at university**. The year 1848 saw the foundation of New England **Female Medical College**, Boston, the first institution of its kind anywhere. The following year, British-born Elizabeth Blackwell became the first woman in modern times to **graduate from a medical school** (Geneva Medical College, now Hobart College, USA). The celebrated English nurse, Florence Nightingale, founded the first modern **nursing school** in 1860. When Mary Edwards Walker worked as a **surgeon** in the Union army (*c.* 1862, USA), she may well have been the first woman openly to hold such a position. Russia's first female military surgeon, Princess Vera Gedroits (1870–1932), went on to become the world's first female **professor of surgery**. Other notable firsts include the graduation in 1864 of Rebecca Lee Crumpler, the first **African American female physician**; the first female to be awarded a **medical degree in India** (1886, Kadambini Ganguly); the opening of the first **female medical school in China** (1902, Hackett Medical College for Women, Guangzhou); the first **female Turkish physician**, Safiye Ali (practising 1922 onwards); and Agnes Savage's graduation from Edinburgh University in 1929, the first **West African woman to qualify as a doctor**.

THE BEGINNINGS OF AESTHETIC MEDICINE

Hindu texts approximately 3,000 years old describe **skin grafting** to restore the features of those mutilated as a punishment for theft or adultery. The first description of purely **cosmetic surgery** (**rhinoplasty** or 'nose jobs', and altered earlobes) is in the Sanskrit *Sushruta Samhita*, the collected works of the Indian sage Sushruta (*c.* 600 BC). Two thousand years later, Gaspare Tagliacozzi (1546–1599) wrote the first **textbook of plastic surgery** (*De Curtorum Chirurgia per Insitionem*, or *The Surgery of Mutilation by Grafting*) for barber-surgeons who were attempting to repair faces ravaged by war, duelling or syphilis. The first **monograph on rhinoplasty** was published in 1845, and in the 1880s the American John Roe pioneered rhinoplasty that did not leave a scar. The earliest **chemical peel**, using a phenol solution, was recorded in 1871. World War I led to several innovations. In 1917, the New Zealander Harold Gillies paved the way for modern **plastic surgery**, and the American artist Anna Ladd developed **anaplastology** by using carefully crafted, wafer-thin galvanized copper to create new faces for gravely mutilated veterans.

A NEW, IMPROVED YOU

Photography and prosperity led to a sharp increase in cosmetic medicine from the late nineteenth century. Autologous (from the same person) fat was first used as a **filler** in 1893 and **paraffin wax** not long afterwards. Around 1898, the German surgeon Vincenz Czerny is recorded as performing a **breast augmentation** (following a patient's cancer treatment), and the earliest mention of a **facelift** comes fourteen years later. After World War II, requests for aesthetic improvement escalated and the number of procedures increased to meet the demand: **hair transplant** (1952), **silicone breast implants** (1961–2), **laser dermatology** (1965), **liposuction** (1977), **gastric bands** (1977), **approved facial fillers** (1981), **breast reduction** surgery (early 1980s), **Botox** (*c.* 1992), **laser hair removal** (1996), **hymenorrhaphy**, **anal bleaching** (both *c.* 2000), and other exotic practices.

MEDICAL HARDWARE

CRUDE (AND PAINFUL) BEGINNINGS

The very first medical instruments were no doubt everyday implements such as knives, put to medical use – from the time of Alexander the Great (356–323 BC) there are reports of a surgeon using a sword to operate. The first three devices most likely to have been created specifically for medical purposes were **tweezers**, **retractors** for holding open incisions (both *c.* 3000 BC, Egypt), and **scalpels** made of obsidian (*c.* 2100 BC, Turkey). We had to wait until 1915 for the first **disposable scalpel** and until 1964 for the **laser scalpel**. Wooden tubes and plant stalks served as **catheters**. By classical times (*c.* 500 BC to *c.* AD 400) there were a variety of metal **hooks**, **spikes**, **drills** and **forceps**, as well as the first **male and female catheters** (S-shaped and straight). Close examination was aided by the convex **magnifying glass** (Roger Bacon, 1250, UK). An Anglo-Saxon hammock-cart may have been the first **ambulance** (*c.* AD 900), though the specially designed vehicles servicing the armies of Revolutionary France from 1793 onwards have a stronger claim. The **obstetric forceps** were probably devised by Peter Chamberlen (*c.* 1560–1631), a French Huguenot refugee living in the UK. In 1752 Benjamin Franklin (see p.151) made the first **flexible catheter** to assist his brother, who suffered from bladder stones. The last significant instrument from the pre-modern era was the **stethoscope**, invented by the Frenchman René Laennec around 1816. With it he was able to examine an obese female patient whose heartbeat was undetectable when he placed his ear to her ample chest.

TECHNOLOGY TO THE FORE

The Dutch surgeon Antonius Mathijsen invented the **plaster of Paris** cast in 1851; an alternative material, **fibreglass**, was introduced from the 1970s onwards. Wilhelm Röntgen's startling discovery of **X-rays** in 1895 (Germany) led to the first X-ray, of a hand, the following year. The procedure was put to **medical use** almost immediately. Thereafter, further inventions came thick and fast: the **electrocardiogram** (**ECG**) in 1903 (Netherlands), the **surgical stapler** in 1908 (Hungary), and the first **human laparoscopy** in 1910 (Sweden). The German psychiatrist Hans Berger recorded the first human **electroencephalogram** in 1924. An Australian anaesthesiologist, Mark Lidwell, invented the **pacemaker** in 1926 and employed it to save the life of a newborn baby in a Sydney hospital. A **portable pacemaker** was made in the USA in 1959. During the German occupation of his country in World War II, the Dutch physician Willem Kolff made the world's first **kidney dialysis machine** (1943) from tin cans and bits of washing machine. Understandably, the prototype did not work particularly well and reliable machines did not become available until after the war.

SCANS AND ROBOTS

The first successful **cardiac defibrillation** – restarting the heart of a fourteen-year-old boy – occurred in Ohio, USA, in 1947. Five years later, punched cards were being used to **correlate data automatically** to help with diagnosis, and General Motors (USA) had built the first **mechanical heart**. By Christmas 1957, doctors were peering inside their patients with a **fibre-optic endoscope**. Another way of seeing what was going on beneath the surface was available the following year with **foetal ultrasound scanning**. Next came new inventions and procedures in cardiac and circulatory medicine: **mechanical replacement heart valves** and the **balloon embolectomy catheters** for removing blood clots (both 1961), **valve replacement** using material harvested from deceased humans (1962), and a Left Ventricular Assist Device (**artificial heart**), which was installed in a patient in 1963. The first commercial **CT scanner** appeared in 1971, the **MRI scanner** in 1977. The 1980s saw **robot-assisted surgery** (1983); the 1990s heralded **stem cell therapy** (1998); and in the first year of the new millennium, **remote surgery** using a robotic system proved practicable. An **artificial pancreas** system for patients with diabetes came on the market in 2017.

BLOOD

The invention of the **microscope** (see p.158) enabled the Dutch naturalist Jan Swammerdam to be the first person to identify **red blood cells** (1658). Almost two centuries later (1841), the compound microscope allowed the British physician George Gulliver to see (and draw pictures of) **platelets**, the colourless cells that cause blood to clot. (Other sources say the official discovery of platelets should be credited to the Frenchman Alfred Donné in 1842.) The term 'platelets' was not used until 1910. In 1843, **white blood cells** (leukocytes) were first observed (simultaneously, apparently) in France by Professor Gabriel Andral and in the UK by William Addison, a discovery that established **haematology**. The German-Jew Paul Ehrlich (who also discovered a cure for syphilis in 1910) took the next major step forward in

1879 with the **staining of tissue** to identify different types of blood cell.

Blood groups (starting with **A**, **B** and **O**) were first identified by the Austrian scientist Karl Landsteiner in 1900. In 1902, two members of his team, the Italians Adriano Sturli and Alfred von Decastello, then identified type **AB**. The **Rhesus (Rh)** blood group system was found – but not named – in 1939. **Rh positive and Rh negative** were distinguished the following year, and in 1967 a **Rh immune globulin** became available to prevent Rh disease in babies born to Rh negative women.

For religious and other reasons, blood – the essence of life itself – was long regarded with superstitious awe. As a result, we have to wait until the early modern era for the first **blood transfusions**. Richard Lower carried out the first reliably documented transfusion (1665, UK), **transferring blood between two dogs**. The transfusion of blood between animals and humans followed (see p.92). The English obstetrician James Blundell carried out the first documented **human to human transfusion** in 1818, though Philip Syng Physick is reputed to have performed the operation in Philadelphia, USA, in 1795. The first **whole blood transfusion** took place in 1840 (UK). The procedure remained hit-and-miss until the discovery of blood groups (see above), the development of **anticoagulants** (notably sodium citrate, 1914), and the introduction of a **citrate-glucose** (1916) enabled blood to be stored for longer periods outside the body. **Blood donation** began in 1921, and the world's first **blood transfusion service** was set up by the British Red Cross in 1926. Five years later, Soviet Russia established the first **blood bank**. Blood collection in **plastic bags** was pioneered in the USA in 1950.

BLOOD OF A LAMB

Jean-Baptiste Denys (1643–1704), personal physician to France's Sun King, Louis XIV, had a thing about blood. The sick could be cured, he believed, if they were given the blood of a healthy animal. His first two experiments went OK: a boy and a man both survived (miraculously!) the infusion of a few ounces of sheep's blood. But the third and fourth operations were less successful. The first killed a Swedish baron and the fourth (late 1667) slew Antoine Mauroy, a poor thirty-four-year-old madman whom Denys had apprehended and given repeated transfusions of lamb's blood. Perrine Mauroy, who had married Antoine during one of his sane periods, had the reckless doctor brought to trial for murder. He was acquitted but gave up medical practice forthwith. Two years later, blood transfusions were banned throughout France.

The earliest reference to what we now know to be **haemophilia** was made in a second-century Jewish document excusing circumcision for boys whose two elder brothers had bled to death after the operation. The first **doctor to record the condition** was Al-Zahrawi (*c.* AD 936–1013), an Arab physician from Cordoba in Spain. Haemophilia's **hereditary nature** was first analysed by the American doctor John Otto in 1803, while the **term haemophilia** (originally haemorrhaphilia) dates from 1828. An **anti-haemophilic globulin** was discovered in 1937 (USA) and the condition's true complexity first noted by an Argentinian doctor who distinguished between **haemophilia A and haemophilia B** in 1947. Judith Pool of Stanford University, USA, discovered cryoprecipitate (1964) as an effective, storable **antihaemophilic factor** (AHF).

VACCINATION, HIV/AIDS AND EBOLA

VACCINATION

By noting that no one caught smallpox twice, the Greek historian Thucydides (*c.* 460 to *c.* 400 BC) was – as far as we know – the **first person to report** on the phenomenon we now call vaccination: administering an infection that stimulates the immune system to withstand a more virulent infection. Questionable sources suggest **variolation** – vaccination against smallpox by deliberately infecting with the disease itself – was **first tried** in tenth-century China, though more reliable evidence comes from the same country in 1549. The practice had reached India, Turkey, Europe and the Americas towards the end of the eighteenth century. **True vaccination** dates from 1797, when the British scientist Edward Jenner made it widely known that deliberate infection with the mild disease cowpox protected against the deadly smallpox. In the next century, Louis Pasteur (see p.94) coined the term 'vaccination' (1891) and developed vaccines against **anthrax** (demonstrated in 1881) and **rabies** (1885).

Louis Pasteur (1822–95), microbiologist and chemist

Since then, many other vaccines have been made, including cholera (1892), typhoid (1896), tuberculosis (BCG, 1921), meningitis (1978) and the sometimes misunderstood MMRV (measles, mumps, rubella and varicella, 2005).

HIV/AIDS AND EBOLA

It is now thought that the human immunodeficiency virus infection (HIV) **passed from monkeys to humans** in about 1910. The first **documented case** of HIV in a human occurred in the Congo in 1959. The term AIDS (acquired immune deficiency syndrome) was **first used** in 1982, when the disease was declared to be an **epidemic** (soon a

pandemic). French scientists **discovered the responsible virus** the following year, and the international community agreed on the term **HIV** in 1986. Effective (but expensive) treatment with **antiretroviral drugs** began in 1997 (USA). Ebola virus disease (EVD) was **discovered** in Africa in 1976, and the first reported case of the disease contracted outside that continent occurred in 2014 (Spain).

SIGHT AND SOUND

SPECTACLES

The ancient Egyptian Ebers Papyrus (*c.* 1500 BC) is said to be the first written work to discuss **diseases of the eye**. The Greek philosopher-physicians Rufus of Ephesus (first century AD) and Galen (see p.61) gave fairly accurate descriptions of the **eye's structure**. (For cataract surgery and an understanding of the physics of sight, see pp.59 and 61.) **Spectacles to aid reading** (convex lenses) are said to date from thirteenth-century Italy, with the first specific reference in 1289. The first person known to have benefitted from **spectacles for short sight** (concave lenses) was Pope Leo X (1517). Spectacles **worn over the ears** appeared in 1727, with **hinges** first added in 1752, while **bifocals** date from around 1785 (USA). Other improvements included the **varifocal** lens (1959, France) and specs with **photochromic** lenses (1965). The earliest description of glasses that **protected from the glare of the sun** comes from twelfth-century China. Lenses that **blocked ultraviolet** light were made in 1913 (UK), while cheap **celluloid** 'shades' appeared in 1929 and **Polaroid** lenses in 1936 (both USA).

TREATMENT AND CONTACTS

John Cunningham Saunders founded the world's first **eye hospital** (now Moorfields) in London, UK, in 1804. **Retinal detachment** was diagnosed the following year. A **chart to measure visual acuity** – the Snellen chart – was created in 1862, giving rise to the American expression 'twenty-twenty vision' (capable of reading a line of letters unaided at a distance of twenty feet (six metres outside the USA). The story of **contact lenses** started with a practical lens made around 1887 (Germany or Switzerland), followed by the **soft** lens (1964), the **bifocal** lens (1982), and the **disposable** lens (1987). In 1916, Germany became the first country to train **guide dogs** for the blind. Photocoagulation for **retinal repair** was written up in 1949, and the first eye **laser surgery** was performed in 1987 (USA). At the time of writing (2019) a number of scientists have developed prototype '**bionic eyes**' – electronic devices wired to the brain which enable the blind to see.

HEARING

The Egyptian Ebers Papyrus (see p.95), groundbreaking in so many fields, was probably the earliest work to discuss **hearing loss**. Although hollowed cow horns were serving as **ear trumpets** by the thirteenth century and purpose-built trumpets were made in the eighteenth century, little was done to aid deafness before the appearance in 1898 of an **electronic hearing aid**, the Akouphone (USA). It was expensive and far too large to be worn on the person – a **wearable hearing aid** was not available until 1935 (USA). Alexander Graham Bell, famous for the invention of the telephone (see p.167), also made the first **audiometer** (1879) to measure a person's hearing, though the first genuinely practicable machine came twenty years later. Hearing-aid technology was revolutionized first with a **transistorized device** (1952) then with a **digital** one (on sale 1987). **Behind-the-ear aids** went on sale in 1989 and **in-the-ear** versions in the 2010s. **Sign language** was devised by the Spanish priest and educator Juan Pablo Bonet in 1620, and **cued speech**, a modern alternative incorporating both signing and lip reading, by the American professor R. Orin Cornett in 1966.

DENTISTRY

PRIMITIVE BEGINNINGS

A form of **dentistry** was practised by our Stone Age forebears. 'Open wide and let's take a look' may have been uttered first by a nimble-fingered Neanderthal over 100,000 years ago, although a crudely excavated molar from Italy (*c.* 12,000 BC) is the earliest concrete evidence. **Dental drills** (*c.* 7000 BC) were spinning in the Harappan civilization of the Indus Valley, and 2,500 years later we have evidence from Slovenia of the first **filling** – made of beeswax! The first **amalgam** filling ('silver paste') comes from China during the Tang Dynasty (*c.* AD 700). The ancient Egyptian Hesy-Ra (see p. 59) is probably the first **named dentist**, and the earliest evidence of an attempt to **drain an abscess** (*c.* 2500 BC) comes from Egypt shortly after his death. At about the same time, the Sumerians were producing some of the first **written accounts of dental problems and treatments**. *Artzney Buchlein* (1530, Germany) was the first **book** (technically speaking, a long section within a complete medical textbook) devoted entirely to dentistry. By 3000 BC, a number of early civilizations were using **chew-sticks** – small lengths of wood with frayed ends – to clean their teeth. The first purpose-made **toothbrush**, using pig bristles, appeared in China in 1498. It is said the ancient Egyptians invented a primitive form of **toothpaste** (*c.* 5000 BC), and **false teeth** and **bridges** (using human and animal extractions) began with the Etruscans of northern Italy in the seventh century BC.

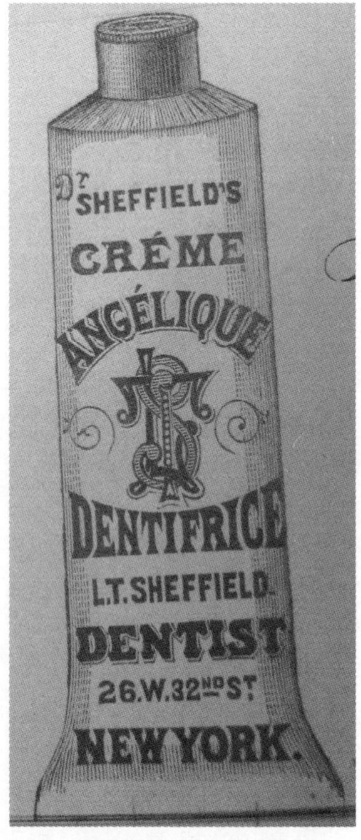

Dr. Sheffield's first toothpaste – *Crème Angèlique*, 1881, USA

MODERN DEVELOPMENTS

Until the lifetime of the physician Pierre Fauchard (the 'Father of Modern Dentistry', 1678–1761), dental practice remained largely a matter of yanking out rotten teeth and occasionally replacing them with staggeringly uncomfortable dentures. The remarkable French pioneer, whose two-volume *Le Chirurgien Dentiste* ('The Dental Surgeon', 1728) revolutionized dentistry, was responsible for a large number of significant firsts. These included **regular dental cleaning by a dentist**, the **link between acid and dental decay** (and overturning the age-old idea that caries is caused by a worm!), inventing the modern **dentist's drill** and **amalgams**, and the **chair light**. Fauchard also instigated modern techniques to help **ease the patient's fears** (such as standing out of sight behind their head). Thereafter, a steady stream of developments led to today's aesthetic dentistry and a relatively comfortable experience in the (in)famous chair: first modern toothbrush (1780), dental school (Baltimore College of Dental Surgery, 1840), toothpaste in a tube (1892), fluoride toothpaste (1950), high-speed drill (1957), and electric toothbrush (on sale 1954).

BIRTH CONTROL AND MENSTRUATION

BIRTH CONTROL

The earliest references to birth control, either through induced **abortion** or **pessaries** inserted into the vagina, are found in Mesopotamian and ancient Egyptian writings from the third millennium BC. The first written reference to ***coitus interruptus*** (*c.* 1400 BC) is the behaviour of Onan in The Bible's Book of Genesis, chapter 38. The first mention of the **death penalty** as a punishment for abortion is in an Assyrian law code *c.* 1075 BC (Iraq/Syria). By the nineteenth century, women were starting to fit purpose-made **diaphragms** (1880s), with interuterine devices (IUDs) obtainable from 1909 onwards. Following years of research and experimentation, the female **birth control pill** became available in 1960; an **implant** version followed in 1983, and the **morning-after** version in 1984 (with a safer version in

2000). The first recorded **vasectomy** was performed in 1823 on a dog, with man undergoing a similar procedure shortly afterwards (both London, UK). A **vacuum pump** to remedy male erectile dysfunction was proposed in 1694 and finally patented in 1913; a more effective drug treatment, **Viagra**, was patented in 1996. Charles Knowlton's coyly named *The Fruits of Philosophy* (1832, USA) was the first widely disseminated **birth control publication**, and the first **birth control clinic** opened in the USA in 1916.

SEX, BIRTH AND BABIES

SEX AND NO SEX

Though not approved of or even accepted in all cultures, **homosexuality** has featured in human society since the species first evolved. Perhaps surprisingly, the word itself did not appear until 1869 (Germany). The word '**asexual**' was first used in 1830, '**bisexual**' in 1824, and '**transsexual**' in 1949. The term '**intersex**' was coined in 1917. '**Gay**' came to mean licentious in the seventeenth century and homosexual (especially male) in the 1920s. In 1890 '**lesbian**' was officially applied to homosexual love between women. The first **sex reassignment surgery** was performed on the Dane Einar Magnus Andreas Wegener, who redefined as Lili Elbe in 1930. The UK's Louise Joy Brown (born 1978) was the first **test-tube baby** (conceived via IVF – in vitro fertilization), and Dolly the sheep, created in 1996 (Scotland, UK), was the first **cloned mammal**.

BIRTH

The earliest examples of **gynaecology**, **midwifery**, **caesarean section** and **forceps delivery** appear elsewhere (see pp.81 and 88). We have a record from 1742 of a birth facilitated by an **episiotomy**, and the first use of **drugs** (chloroform) to ease the pain of childbirth was in 1847. An **epidural anaesthetic** was administered in Spain in 1921. Long ago, **water births** are said to have been practised in ancient Egypt and the South Pacific, but the first documented example comes from France in 1805.

INFANTS

The ancient Egyptians (who else?) probably pioneered **non-human baby milk** and **feeding bottles** some 3,500 years ago. Modern **formula milk** for babies dates from 1867 and was first available as a **powder** in 1915 (see p.25). Bespoke **baby foods** went on sale in the Netherlands in 1901. No doubt, objects were placed in babies' mouths to comfort them from time immemorial, though the first mention of a **dummy** (or 'pacifier') in literature was not made until 1473 (Germany). Wooden **cribs** are as old as civilization, with the first rockable iron **cot** with sides to stop the infant falling out appearing around 1630 (UK). The **playpen** dates from the 1880s, the **baby jumper** from 1910 (or earlier), and the seat-like **baby bouncer** from 1961. The first **pram** – a true 'baby carriage' – was built for the Duke of Devonshire in 1733 (UK), with an aluminium **folding version** following two centuries later (1965, USA). The first disposable nappies have been noted elsewhere (see p.57) which leaves only the **babygrow** (Babygro, 1950s, USA) to complete our baby's trousseau.

Interestingly, **gender colour coding** is relatively new: white was the norm for both sexes until the late nineteenth century; in the early twentieth century it was pink for boys and blue for girls; not until the 1940s did this reverse to the present norm.

PART IV: GETTING ABOUT

HORSES AND CARTS

TAMING THE HORSE

Oxen, the original **draught animals** (or beasts of burden), were working for their human controllers perhaps 10,000 years ago (see p.164). There is evidence to suggest that while the people of the Fertile Crescent and Eastern Europe were trundling about in the first **carts** (*c.* 3200 BC, see p.104), the inhabitants of the Kazakhstan region were **domesticating horses** either as food or as a means of transport. Another millennium passes before the Bronze Age site at Chelyabinsk Oblast, Russia, provides sure evidence of horses as draught animals. It also gives us the first **bridles** and **bits** – but no saddles. The Assyrians 'came down like a wolf on the fold' with fringed cloths between themselves and their steeds *c.* 700 BC (Iraq/Syria). Shortly afterwards, they and their neighbours were benefiting from **saddles**. Indians invented the **stirrup** in the second century BC. The modern, more secure version – that allowed the likes of Sir Lancelot to remain in the saddle – dates from China around AD 300.

WHEELS AND CARTS

The first **wheels** were used to make **pottery** (*c.* 3500 BC – see p.145). Apparently, it took about 300 years of watching potters' wheels rotate before some genius in Mesopotamia or Eastern Europe had the bright idea of turning two of them on their side and attaching them to a platform to make a **cart**. While it's impossible to ascertain precisely when this happened, we can identify the oldest extant example of a **wheel for transport**: recovered from marshes near Ljubljana City (Slovenia), the ash–oak construction is thought to be approximately 5,200 years old. Though an even older sketch of what may be a four-wheeled vehicle has been found in Poland, the first undisputable images of battle carts or **chariots** are on a 4,500-year-old wooden Sumerian box known as the Standard of Ur. Wooden **spoked wheels** were first made around 500 years later in Siberia; *c.* 500 BC, the Celts were the first to add iron rims. Next came three significant firsts

Sumerian battle wagons on the Standard of Ur, c. 2500 BC

attributed to Roman engineers (late first century BC): **metal wheel bearings**, **sprung suspension** (using chains or leather) and the **pivoted front axle**. The Romans are also credited with inventing a foot-operated **brake**.

CARRIAGES

The animal-drawn cart remained little changed until the nineteenth century. However, the light and fast people-carrying **carriage** (or **coach**, a name derived from the Hungarian town of Koch where they were first made in the mid-fifteenth century) went through many variations. The two-wheeled, one- or two-person **chaise** was bowling along the streets of Western Europe towards the end of the seventeenth century. Roughly thirty years later, it was joined by the low-slung and luxurious **landau**, first manufactured in the German city of that name. In the mid-eighteenth century, the French made the one-horse **cabriolet**, soon shortened to 'cab' and made famous by Joseph Hansom's **hansom cab** (1834, UK); it's still with us as the 'taxi cab'. The **barouche**, a smaller version of the landau, appeared at the turn of the nineteenth century. By this time, **steel springs** had been in use for about 150 years. The first **stagecoach** (1610) was Scottish, running between Edinburgh and Leith, and the first **mail coach** clattered between London and Bristol in the UK (1782). A horse-drawn **bus service** was initiated in London in 1829.

BEASTS OF BURDEN

The long-suffering ox (see p.164) was followed on the beasts-of-burden timeline by **llamas** (in the vicinity of Lake Titicaca in the Andes, between Bolivia and Peru) and **donkeys** (aka **asses**, ancient Egypt), both first employed for carriage some 6,000 years ago. **Dromedaries** (one-humped camels) were carrying loads across the arid areas of southern Arabia *c.* 3000 BC, and about 500 years later two-humped **camels** were doing the same in the region of Persia (Iran). By this time, the Egyptians were engineering what had been going on in the wild for ages – crossing male donkeys with female horses to produce load-carrying **mules**. The Indians tamed **elephants** *c.* 2000 BC. The nineteenth-century development of the European **dog cart** may refer to a small vehicle pulled by one or more dogs, or a light, horse-drawn chaise that carried hunters and their hounds. Human beasts of burden carried the first **litters** (*c.* 4000 BC, ancient Egypt) and the enclosed, hireable **sedan chairs** (late sixteenth century, France). Early civilizations may have moved heavy loads on **sledges**; nevertheless, the earliest extant example of a sledge is in a late eighth century ship uncovered in Oseberg, Norway.

THE LONG ROAD TO THE AUTOMOBILE

SELF-PROPULSION

The earliest known design (1478) for a **self-propelled vehicle** (or auto mobile) is that of the Italian genius Leonardo da Vinci. Though his sketchy plans are unclear, scholars now believe his car was to be spring-powered (i.e. clockwork). Next (*c.* 1672) came a **steam-powered** toy vehicle designed by a Jesuit missionary for the eighteen-year-old Chinese emperor Xuanye. In 1769, Frenchman Nicolas-Joseph Cugnot built a

Karl Benz driving with his business partner Max Rose as passenger, 1887

working steam-powered, three-wheeled military **tractor**, and at the start of the following century, Cornishman Richard Trevithick demonstrated his *Puffing Devil* **road carriage** (1801). In 1807, Switzerland became the birthplace of the first vehicle powered by an **internal combustion engine** (hydrogen-fuelled). It was not practicable. Nor were early **electric vehicles**, a concept apparently conceived by the Scotsman Robert Anderson in the 1830s. The horseless carriage of the Belgian inventor Jean Joseph Étienne Lenoir featured the first **commercially viable internal combustion engine** (1863) and a **petrol-driven** version was built in Vienna, Austria, in 1870. Seven years later, the German engineer Nikolaus Otto developed the first **four-stroke engine**. All this led to Karl Benz's construction of what is widely regarded as the first **modern car** (1885, Germany).

BUBBLE CARS

In the aftermath of World War II, Germany's shattered aircraft industry had engineers and imagination but little capital. The domestic market was strapped for cash, too. Designers at Messerschmitt, seeking to exploit their wartime know-how (especially of engines and cockpit canopies), put their heads together with an invalid carriage manufacturer and came up with . . . the KR175 bubble car (1953). Other companies followed suit. The souped-up carriages enjoyed short-term success until superseded by the Mini (1959, UK), whose concept of a transverse engine mounting had been pioneered in 1911. In 1998, the small-is-beautiful mantra passed to the Smart microcar (1998, France).

MASS PRODUCTION

Benz's 1885 car was a three-wheeler. The first modern **four-wheeler** was the Cannstatt-Daimler of 1886 (Germany). The sixty-horsepower Mercedes of 1903 (Germany) called itself a fast **tourer**, but the accolade of first **sports car** (a phrase not actually used until after World War I) is generally awarded to the three-litre Vauxhall Prince Henry (1910, UK).

The Dutch Spyker 60 HP introduced **four-wheel drive** in 1903. Though the famous Model T Ford was the first car made on a **moving assembly line** (1913, USA), the first **mass-produced** car was the 1901 Oldsmobile (USA; see also p.111). Early vehicles were all open-top. Then came **limousines**, where the enclosed passengers were separated from the driver (who wore a cloak like those worn in the French region of Limousin), and **stretch limos** (1920s, USA). Roofed cars paved the way for the **convertible** (1922) and the **power-operated retractable top** (1934). By this time, a number of unusual variations on the conventional automobile were on the market: the DKW Typ P (1928, Germany), powered by a **two-stroke** engine; the Zaschka, a three-wheel **folding car** (1929, Germany); and non-folding **three-wheelers** in the form of Germany's Goliath Pioneer and Britain's Raleigh (later Reliant) Safety Seven (both *c.* 1931). **Diesel**-car production began with France's 1933 Citroën Rosalie, while the US army's Jeep (1940) launched the **4x4 all-purpose** vehicle. The **piston-less rotary** (or Wankel) **engine** first ran in 1957 (Germany; see also p.150). After a new wave of prototype **electric cars** in the 1990s (e.g. the GM EV1, Honda EV Plus and Toyota RAV4 EV), we had to wait until 2008 for the first viable model: the Tesla Roadster (USA). Eleven years earlier, Toyota had launched the Prius, the first mass-produced **hybrid vehicle** (Japan). In 1977, a Japanese laboratory produced a type of **driverless car**; in 2017, Audi (Germany) claimed its new A8 would be the first **fully automated car** – but only to the speed of 60 kph.

PERMISSION TO DRIVE

By 1888, the noise and fumes of Karl Benz's Motorwagen (particularly after his wife's 100-km dash from Mannheim to Pforzheim on 5 August – the first long-distance car journey) were causing enough complaints for local authorities to insist he obtain written permission to drive on public roads – the world's first driving licence. Number plates (1893, Paris, France) and a mandatory driving test (1899, France) followed. The UK introduced mandatory licences for all drivers in 1903.

GADGETRY GALORE

Despite a car with a **steering wheel** taking to the road as early as 1894 in France, the steering tiller (like a ship) remained the norm for several more years. Uncertain handling and braking led to the need for effective **bumpers**, first fitted in 1901 (UK). In 1910, Frenchman Louis Renault made driving a little less precarious when he built on the work of the German engineer Gottlieb Daimler to produce the **drum brake**. Racing cars constructed by the German-born American Fred Duesenberg in 1914 featured the first **hydraulic brakes**. The first **disc brakes** were fitted in 1902 (Frederick Lanchester, UK), although they were not widely used until much later, and **ABS** (invented in 1929, France) initially featured on a Jensen Interceptor (UK) in 1966. **Differential drives** were employed on the early Benz cars (see p.108). The first **clutch** (1893, USA) led to the **gearbox**

(1894, France) and the replacement of the chain by a **shaft drive** in 1901 (USA). Cadillac (USA) pioneered the **starter motor** (1912) and **synchromesh** gearbox (1928). Dynamo-generated **electric light** was installed in a car as early as 1901 (UK) – a luxury 'extra' that doubled the cost of the vehicle; by 1922 dynamos were sufficiently powerful to run a **car radio** (Chevrolet, USA).

SAFETY AND THE ENVIRONMENT

The first recorded **road traffic fatality** to involve a motorized vehicle occurred in Ireland, when the unfortunate Mary Ward was run over and killed by a steam car in 1869. With the advent of the motor car, the death toll mounted rapidly. **Grooved tyres** (1908) had only a minimal impact on safety. **Safety glass** (1909) and **tubeless tyres** (1946) helped a little, as did **seat and fascia padding** and **recessing of sharp edges** (1937 onwards). Saab (Sweden) devised the **safety cage** in 1949, and another Swedish manufacturer, Volvo, introduced the three-point **seat belt** (1958), making it a standard fitting the following year. Mercedes-Benz (Germany) pioneered the **crumple zone** in 1959, and the Oldsmobile Toronado (USA) was the first car on general sale with a **passenger airbag** (1973). The health of those outside the vehicle first received attention with compulsory **emission standards** (1966, California, USA), and the invention of the **catalytic converter** (patented 1956, into production 1973, USA). The previous year, Japan was selling **unleaded petrol**; it **banned leaded fuel** in 1986.

OTHER ROAD USERS

TAXIS AND BUSES

The first **self-powered taxis** were probably the electric 'Hummingbirds' that flitted through the streets of London in the summer of 1897 (UK). The **meter**, a German invention, had appeared six years earlier; the Daimler Victoria (Germany) was the first **petrol-driven taxi** to have one fitted (also 1897). They remained mechanical until the 1980s when electronic versions became available. A Daimler Victoria was converted into the first **motor bus** in 1895, and its successors rapidly replaced existing omnibuses: **horse-drawn** (first seen in Paris, France, in 1823) and **steam-powered** (chugging around the UK from the 1830s onwards). Paris saw the first horse-drawn **double-decker bus** (1853) and the first **double-decker motor bus** (1906, France). A prototype **articulated bus** was dreamed up in Budapest in the 1920s (Hungary). '**Charabanc**' – a bus for pleasure outings, derived from the French *char-à-bancs* (carriage with benches) – took its first load of happy day-trippers in early twentieth-century Britain.

LORRIES, TRACTORS AND TRACKS

Karl Benz, the remarkable motor car pioneer, built the first petrol-driven **motor lorry** (or truck) in 1895. Surprisingly, a **diesel lorry** did not follow until 1923 (Benz again!). **Articulated vehicles** had been around since 1881. **Containers**, used for shipping in the late eighteenth century, were adopted by the railways in the 1830s and standardized by the US army in World War II. The modern **standard steel container** dates from 1956 (USA). Steam-powered **traction**

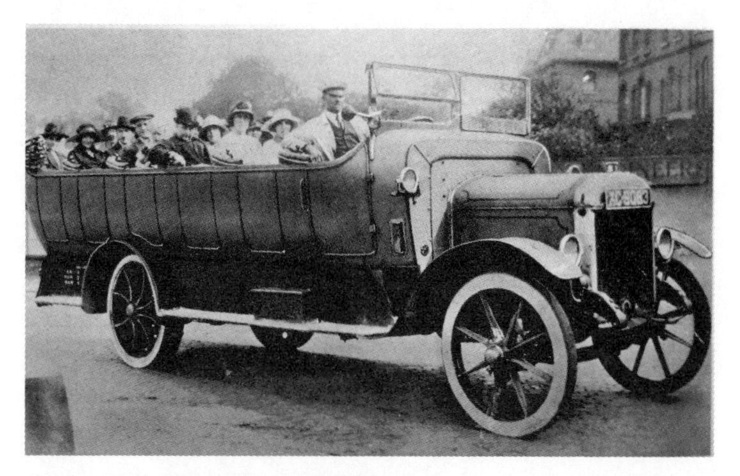

A motorized Charabanc of the 1920s

engines were launched in 1859 when Thomas Aveling fitted a stationary engine with a drive shaft (UK). John Froelich built the first **motor tractor** in 1892 (USA). The theory of **caterpillar tracks** ('a universal railway' or 'endless rails') was worked on in the 1830s by Józef Maria Hoene-Wroński (Poland), Sir George Cayley (UK) and Dmitry Zagryazhsky (Russia), but the first commercial **crawler tractor** (steam-powered) did not go on sale until 1901 (the Lombard Steam Log Hauler, USA). A **motor-driven crawler** was made in 1908 (UK), and motor-driven **snow vehicles**, featuring tracks (and sometimes skis), originated in Canada in 1912. **Self-propelled cranes** were available on railways from 1886 and on roads from the 1910s.

BIKES...

Few areas of human ingenuity and invention are more controversial than the humble **bicycle**. Amid all the claims and counter-claims, I suggest the following firsts are more or less accurate. In 1817, Baron Karl von Drais (Germany) invented a steerable, two-wheeled 'running machine' (aka the **velocipede** or hobby-horse) – the earliest bicycle. Around 1869, the first metal-framed, pedal-powered bicycle (the '**boneshaker**') went on sale in France. In 1871, the British engineer James Starley invented the high bicycle or '**penny-farthing**'. In 1885, John Kemp Starley (James's nephew) invented the Rover or '**safety bicycle**' – the prototype for all modern bikes. The **diamond frame** (1889) and **gears** (1900s) soon followed.

... AND MOTORBIKES

In 1867, Ernest Michaux, son of the French boneshaker pioneer Pierre Michaux, fitted a small steam engine to one of his father's machines – producing the first **motorbicycle**. Sort of. After many designers had tried their hand at marrying the bicycle with the internal combustion engine, in 1894 Hildebrand & Wolfmüller (Germany) manufactured a commercially available *Motorrad* or **motorcycle**, thereby launching both the product and the name. The world-renowned **Harley-Davidson** was founded in 1903 (USA). The **sidecar**, a British innovation, dates from the same year.

THE HIGHWAY

ROADS

The earliest constructed **streets** were probably those of the cities of ancient Mesopotamia (*c.* 4000 BC). British Celts were constructing **log roads** about the same time. The first **paved roads**, or highways, were built by the ancient Egyptians some 1,500 years later. Flattened and compounded road surfaces were greatly facilitated by the arrival of the **steamroller**, a French idea (1860) that went into production in the UK in 1867. Perkins (UK) began making **motor-driven road rollers** in 1904. The favoured modern paving, **tarmac**, was invented by accident in 1902 (UK). Charging a **toll** for road use goes back to the reign of Ashurbanipal (668 to *c.* 627 BC), the last effective king of Assyria, who charged travellers journeying between Susa and Babylon (Iraq/Syria). **Electronic toll** collection, pioneered in the USA, was first put into large-scale operation in Bergen, Norway, in 1986.

SIGNS AND PARKING

The first **road signs** were likely to have been the **milestones** placed on Rome's Appian Way (312 BC). After that, road signs were largely a matter of local whim. In 1697, the English parliament empowered magistrates to position directional **fingerposts**, and made them obligatory on turnpikes in 1766. King Peter II of Portugal instigated **priority signs** on the narrowest streets in his capital city in 1686. Signposts, together with **street** and **town name** signs, became widespread towards the end of the nineteenth century, as did **hazard warning signs** erected by intrepid cyclists. The Italian Touring Club introduced the first **comprehensive road sign system** in

1895 for its newly united country; an **international system** was agreed at the 1908 International Road Congress in Paris, France. The UK painted the first **white lines** on roads in 1918 and installed the first **cat's eyes** (patented by Percy Shaw, 1934). Designated **car parks** appeared in France and the US around 1900, and in 1905 the first **multi-storey, automated car park** was built in Paris. Oklahoma City (USA) launched the **parking meter** thirty years later.

ACCIDENTS

Though Karl Benz's 1885 Motorwagen (see p.110) ran into a wall during a demonstration run, the first genuine **motor car accident** was probably when James Lambert's vehicle hit a tree root and crashed into a hitching post (1891, Ohio City, USA). Five years later, the British **pedestrian** Bridget Driscoll was struck and **killed** by a motor car travelling at 'a reckless pace' (4 mph) beside the Crystal Palace, London (see p.160). Street crossings in the form of raised stepping stones existed in Roman Pompeii (first century AD), and the first designated **pedestrian crossing** was established on a London (UK) street in 1868. Crossings were marked by **Belisha beacons** in 1934 (UK), and painted with stripes (the **zebra crossing**) in 1951 (UK). Over the centuries, numerous countries tried to curb 'furious driving' by imposing restrictions on drivers, making the first **speed limit** impossible to ascertain. However, we do know the first car driver to be **fined for breaking the speed limit**: in 1896, the Englishman Walter Arnold had to pay one shilling (twelve old pence) for travelling at 8 mph. **Speed cameras**, a Dutch invention, were first installed around 1968 and **radar speed checks** in 1971 (USA). Compulsory **car insurance** was introduced in the UK in 1930.

CROSSINGS

Semaphore traffic control began in 1868 (UK) and **traffic lights** date from 1914 (USA). Bath's classical-style Circus (1768, UK) may be the first **circular traffic junction**, though the first **roundabout** designed specifically to keep the traffic flowing was built in San Jose, California (USA) in 1907. The Greek Mycenaean civilization (*c.* 1300–1190 BC) probably produced the earliest stone **road bridges**. The first iron bridge opened in 1781 (UK), while the first **suspension bridge for road vehicles** was Jacob's Creek Bridge, Pennsylvania (1801, USA), and by the 1860s the French were building **concrete road bridges**. The Via Flaminia's tunnel under the Apennines at Furlo Pass, Italy, was the first **road tunnel** (AD 77); London's Thames Tunnel was the world's first **underwater road tunnel** (1843, UK), and Holborn Viaduct was the first **road flyover** (1869, UK). **Motorway** construction started with the Long Island Motor Parkway (1907, USA).

MAKING TRACKS

THE RAILWAY

The ancient Athenians ran **wheeled vehicles in stone grooves** (*c.* 500 BC); the Austrians hauled trucks up **wooden rails** in 1515; and the British introduced **iron-capped rails** in the 1760s. **Flanged wheels** made their appearance in the same year as the French Revolution (1789), and **wrought iron rails** soon afterwards (UK). The next step was to replace horse and human power with steam locomotion. Richard Trevithick (see p.107) constructed the first **steam locomotive** in 1804. A steam-powered **rack railway** followed (1812, Leeds, UK), then ***Puffing Billy***, the first steam locomotive able to **propel itself smoothly and continuously** along metal rails. A public **part-steam railway** opened between Stockton and Darlington (UK) in 1828, and the following year the Liverpool–Manchester line (UK) became the first in the world to operate by **steam power alone**.

TRAINS AND TRAMS

The first **battery-powered locomotive**, built in 1837 by Scotsman Robert Davidson (UK), could not compete with its steam counterparts. Over forty years passed before Werner von Siemens overcame the battery problem by constructing an **electric locomotive** that took its **power from a rail** (1879, Germany); in 1895 the first **electrified main line** went into service in the US. **Trams** spread rapidly after the first rattled along its Welsh tracks in 1807 (UK), with electricity swiftly replacing horse and steam power after Russia had pioneered the **electric tram** (streetcar) in 1880. Early experiments with petrol- and diesel-powered locomotives (1888 and 1894, UK) made little mark, and the first **diesel-powered locomotive** did not enter service until 1912 (Switzerland). A **diesel-electric** railcar ran in Germany two years later and was commercially much more successful. Japan's famous Shinkansen 'Bullet Train', inaugurated in 1964, was the world's first regular **high-speed** (210 kph / 130 mph) railway. The London **Underground** ('the Tube'), opened in 1863 (UK), was another first. So too was its **electrified line**, the City & South London Railway, which opened in 1890. The earliest **elevated railway** ran along the 878 brick arches of the London and Greenwich Railway

Departing ceremony of a Tokaido Shinkanses bullet train, Tokyo Station, 1964

(1838, UK); the Liverpool Overhead Railway (UK) was the first powered exclusively by **electricity**. The Russian inventor Ivan Elmanov designed the first **monorail** in 1820. A wide variety of designs, prototypes and small-scale tracks followed before the **wide beam** ALWEG Monorail (1957, Seattle, USA) finally proved the concept practicable and commercially viable.

DEATH ON THE LINE

The earliest recorded railway accident occurred when two County Durham boys were run down on a coal tramway in 1650 (UK). Reports of further accidents were fortunately few until 31 July 1815. On that summer morning, shortly after the Battle of Waterloo, the Derbyshire engineer William Brunton was demonstrating his extraordinary Steam Horse (aka the Mechanical Traveller) to a large and fascinated crowd gathered at Philadelphia, also in County Durham. Pushed by two steam-driven metal legs, the Horse was plodding along at about 3 mph when its boiler exploded. Sixteen people died in the blast, the first major railway disaster. Twenty-seven years passed before this shocking death toll was exceeded: in 1842, a French train with its passengers locked inside its carriages crashed and caught fire. The exact number of deaths was impossible to ascertain, but it may have been as many as 200.

TICKETS AND TIMETABLES

The Mount, on Wales's horse-drawn Oystermouth Railway, was the first railway (or **tram**) **station** (1807), and Liverpool's Crown Street Station the first **terminus** (1830, UK). As mechanical **signals** replaced hand signals in the 1830s, a central **signal box** became feasible (the first opened on the London and Croydon Railway in 1843, UK). Early handwritten **tickets** were replaced by Thomas Edmondson's printed ones (cut in half for **half-fares**) from 1839, the same year as the earliest printed **timetable** appeared (both UK). **Season tickets** (hand-written) were launched on the tiny Canterbury and Whitstable Railway in 1834 (UK). *Experiment*, the first purpose-built **railway coach**, was constructed in 1825 (UK), with **sleeper cars** available from 1839 and dining cars from around 1866 (both USA). The first **international railway service** (1842) ran between France and Belgium, nine years after the first **train ferry** had sailed in Scotland, UK.

ON WATER

BOATS, ROPE AND SAILS

Suggestions for the earliest boat range from 10,000 to 900,000 years ago! Some say the first vessels were **coracles** made of animal skins stretched over wooden frames; others reckon the **dugout canoe** takes precedence. Whichever came first, both were steered and probably driven by some form of **paddle** or **oar**. The 10,000-year-old remains of a dugout have been found in the Netherlands; evidence from the first **raft** indicates it was made 2,000 years after this (Egypt); the Sumerians had **sailing boats** as early as 6000 BC. Three thousand years later, the ancient Egyptians were making the first **ocean-going vessels** with hulls of joined **planks**. The toothed **anchor** appears to date from about 1000 BC, while the sternpost **rudder** was a Chinese invention of the first century AD. Simple **rope** was made at least 28,000 thousand years ago, but the Egyptians were the first to make strong, **multi-strand rope** from reeds and other fibres (*c.* 4000 BC). The first record of a ram on a ship – making it arguably the first **warship** – was in 535 BC (Greece).

Notable other firsts include **galleys** (*c.* 700 BC), **galleons** (sixteenth century), the purpose-built **lifeboat** (1790), **liner** (1840), **tanker** (1878), **cruise ship** (1900), Dreadnought **battleship** (1906), purpose-built **aircraft carrier** (1922), and **container ship** (1955).

CANALS

The Sumerians of ancient Mesopotamia (Iraq) pioneered **irrigation canals** around 3500 BC. The first **navigable canal** may have been that completed *c.* 510 BC by Darius I of Persia to link the River Nile to the Red Sea. Nevertheless, the evidence for China's Grand Canal, begun in the third century BC, is more reliable. The Chinese engineer Chiao Wei-yo designed and constructed the first modern **lock** (AD 984). The Suez Canal (opened 1869, Egypt) was the first major **ship canal**.

IRON, STEAM AND BEYOND

The launching of the French *Pyroscaphe* in 1783 was a double first: a **steam-powered ship** and a **paddle steamer**. (The idea of a ship driven by rotating paddles, however, goes back to Roman times.) The *Charlotte Dundas* (1803, Glasgow, UK) was the earliest **practical steamboat**. A river barge of 1787 was the first **metal-hulled boat**, an **iron passenger vessel** following in 1819 (both UK). The *Aaron Manby*, launched in London (UK) in 1822, marked another double first: an **ocean-going iron ship** and a **steam-powered iron ship**. *Gloire* (launched 1859, France) was the first **ironclad warship**. In 1807, the world's first **internal combustion engine** was installed on a boat (France, see p.149), though one powered by a modern **petrol engine** did not appear until 1886 (Germany). The French submarine *Aigrette*, the first **diesel-powered ship**, took to sea in 1905. **Turbine engines** were invented in 1791 (John Barber, UK), and the experimental, high-speed *Turbinia* of 1894 (UK) paved the way for **turbine-powered ships**. Finally, the submarine USS *Nautilus* (1955) and the Soviet icebreaker *Lenin* (1959) were the first **nuclear-powered ships**.

SPORT ON THE WATER (see also p.255)

After **yachting** – sailing for fun – was established in the mid-seventeenth century (see p.253), its rising popularity led to the foundation in 1720 of the first **yacht club**, in Cork, Ireland. The Royal Thames Yacht Club (UK) claims its 1775 race for the Cumberland Cup was the first proper **regatta**. The first set of **international specifications** was drawn up in 1906 (International Conference on Yacht Measurement, London, UK). **Dinghies** – pleasure craft for the less well-off – are said to have been pioneered in 1887 by the Irish solicitor Thomas Middleton. The modern **surfboard** dates from 1926 (Tom Blake, USA), **windsurfing** from 1958, and **kiteboarding** from 1977 or just before (both USA). **Canoes** and **kayaks** were among the first boats ever constructed (see p.121). The flat-bottomed **punt**, originally developed in the Middle Ages for navigating the shallow waters of East Anglia's fens, was first used as a pleasure craft on the River Thames in the 1860s (UK). The British naval officer Lieutenant Peter Halkett invented the **inflatable rubber boat** in 1844, and in 1907 a Norwegian inventor came up with an **outboard motor** to power it. The **personal watercraft**, or **Jet Ski** (USA), first began shattering the peace of the seaside in 1972.

SCREWY THINKING

The concept of a propeller (technically a screw propeller) originated with the Greek mathematician Archimedes (*c.* 287 to *c.* 212 BC) who invented the screw for lifting water. The idea of reversing the process – using the motion of the screw to move it through stationary water – took a remarkably long time to materialize. It eventually came into being out of necessity. For obvious reasons, David Bushnell's *Turtle* (1775, USA), one of the earliest submarines, could not sail with the wind; nor could it be rowed or paddled like a surface ship. Bushnell responded by fitting his craft with a hand- or foot-powered propeller. Sixty-three years later, SS *Archimedes* was launched: the world's first steamship driven by a screw propeller (UK). The first self-propelled submarine was the French *Plongeur* (1863).

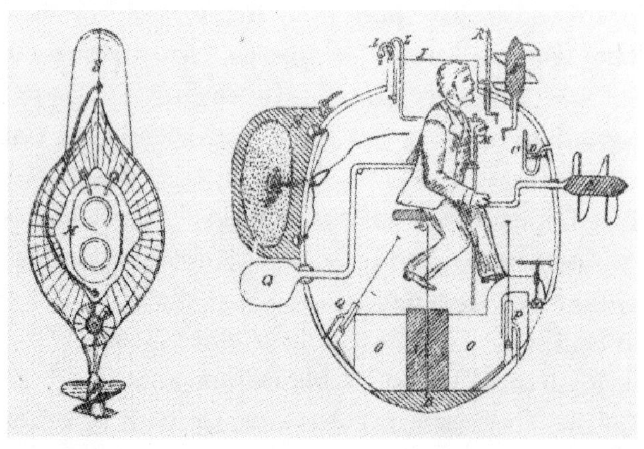

The design of Bushnell's *Turtle*, an early submarine, 1880–1890

UP, UP AND AWAY

KITES AND BALLOONS

The **kite** is a Chinese invention. It may date from the fifth century BC, and kites were certainly flying by AD 550. The earliest mention of a **human kite flight** is the supposed 2.5 km journey of Prince Yuan Huangtou in AD 559. The Chinese also pioneered unmanned **hot-air balloons** around the start of the first millennium AD. In 1709, the Portuguese Jesuit Bartolomeu de Gusmão may have lifted himself off the ground with a form of hot-air balloon, but the ten-minute flight of the French Montgolfier brothers (1783) is generally recognized as the first human **journey by balloon**. The first **hydrogen balloon** had taken off in France a few days before, and one large enough **to carry people** lifted off in Paris on 1 December – making 1783 a true annus mirabilis in the ballooning almanac. The crash of a hot-air balloon on the town of Tullamore, Ireland, destroying a hundred homes, was the first **aviation disaster** (1785). **Helium**, rather than highly combustible hydrogen or coal gas, was first used in World War I barrage balloons (1917–18, USA) and in an airship in 1921 (USA). The history of the dirigible balloon or **airship** began when Frenchman Jean-Pierre Blanchard equipped his balloon with a hand-powered propeller in 1784. Another Frenchman, Henri Giffard, prepared the way for **engine-powered airships** when he fitted a steam engine to his craft in 1852. The German aviator Paul Haenlein flew an airship with an **internal combustion engine** in 1872; the first of the most famous airship marques of all, the German **Zeppelin**, became airborne in 1900.

HEAVIER THAN AIR

Sir George Cayley (UK) pioneered heavier-than-air flight. He was the first to work out the **principles of aeronautics**, build a working **glider** (1804), and fly a **glider carrying a full-grown person** (1853). In 1890, Frenchman Clément Ader's steam-powered flying machine **flew uncontrolled** for 50 m (160 ft). The first **controlled flight by a powered heavier-then-air machine** is generally ascribed to Orville Wright (USA) in December 1903, but a good case can be made for an earlier flight (August 1901) by the German-born Gustave Whitehead (USA). Once people had taken to the air, progress was rapid. In 1906, the Romanian Trajan Vuia constructed the first **monoplane**. Two years later, Thérèse Peltier (France) became the first **woman to fly solo**, and her compatriot Louis Blériot flew the English Channel in 1909. In 1910 there were two more significant firsts: a **seaplane** (France), and **take-off from** and **landing on a ship** (USA). Aeronautical display took a spectacular turn when the Russian pilot Pyotr Nesterov **looped the loop** in 1913.

The Wright Brothers' famous first flight at Kittyhawk, 1903, USA

LARGER AND FASTER

January 1914 saw the initial flight of the four-engine Sikorsky Ilya Muromets (see p.57), the pioneer **airliner** and the first **four-engine plane** – and a prototype version of the same craft had probably been the first **two-engine plane**. The Russian giant never flew commercially because of the outbreak of World War I, leaving the St Petersburg–Tampa Airboat Line to become the world's first **scheduled airline** (1914, USA). The war brought rapid aircraft development, including the **all-metal** Junkers J 1 (1915, Germany). The first regular **international passenger service** opened between France and Belgium almost immediately after hostilities ended (1919). **In-flight meals** were served shortly afterwards (UK) and **in-flight movies** were first shown in 1921 (USA). **Air refuelling** started in 1923 (USA). Frank Whittle (UK) patented the **jet engine** in 1930, but the first **jet aircraft**, the German Heinkel He 178 V1, did not fly until 1939. Jet power enabled the **sound barrier** to be broken in level flight in 1947 (USA); the first **jet airliner**, the de Havilland Comet (UK), made its first commercial flight five years later, and the Anglo-French Concorde, the first **supersonic airliner**, took off in 1969. In 1974, an entirely new form of propulsion emerged with the **solar-powered** Sunrise 1 (USA).

UNUSUAL CRAFT AND EXITS

The advance of aeronautical science and technology gave rise to a variety of aircraft. The story of the **helicopter** (a word coined in 1861) goes back to bamboo Chinese tops (*c.* 400 BC), but does not really begin until the nineteenth century when several **models** and **unmanned craft** wobbled into the air, powered by steam, electricity and petrol. At last, on 13 November 1907, the Frenchman Paul Cornu devised a **stable, piloted helicopter** that rose 30 cm and hovered there for twenty seconds. A Spanish **autogyro** flew in Spain in 1923, and in 1944 the Sikorsky R-4 (USA) became the first **mass-produced** helicopter. The quest for a true **VTOL** (Vertical Take Off and Landing) plane went through various phases before the jet-powered Short SC.1 flew in Belfast (Northern Ireland, UK) in 1957. **Hovercraft** began with the flight of SR.N1 (UK) two years later. **Hang gliding** goes back to the late nineteenth century, though the modern design dates from 1961. **Microlights** took off in the 1970s; modern **drones** (unmanned aerial vehicles) took to the skies with the RP-1 (1935, USA), and **battery-powered quadcopter drones** became commercially available *c.* 2010. The **parachute** was demonstrated in France with a jump from a tower in 1783; the first jump from an aircraft was in 1911 (USA). **Ejector seats**, a Romanian invention, were tested successfully in 1929. The world's first **airfield** was supposedly near Viry-Châtillon, France, and the first **airport** opened at Hounslow Heath, near London, UK, in 1919.

TO THE MOON . . . AND BEYOND

The first reported deployment of **rockets** was by the Chinese in 1232. They remained occasional weapons of war until Konstantin Tsiolkovsky, the 'Father of Modern Astronautics', proposed space exploration using **liquid fuel** in 1898. The first such **rocket flew** in 1926, and in 1944 a German V2 rocket **reached space**. Thirteen years later, the Russians surprised the world with Sputnik 1, the first man-made **Earth-orbiting satellite**. The US had launched a **monkey**

Neil Armstrong, the first man on the moon, 20 July 1969

(Albert 1) into space on a V2 rocket in 1948, and the USSR put a **dog** (Laika) into **orbit** in 1957. In 1961, Yuri Gagarin became the first **human being in space**, followed in 1963 by Valentina Tereshkova, the first **woman in space**. A Russian man made the first **spacewalk** in 1965, the same year as the US pioneered the **space rendezvous** (docking). On 20 July 1969, Neil Armstrong and Buzz Aldrin of the US Apollo 11 mission **walked on the moon**. Another fifty years passed before the Chinese landed a spacecraft on the **dark side of the moon**.

A **soft landing** on Venus was accomplished in 1970, followed by one on Mars the next year (both Russia). In 1971, the Russians also established the first **space station**. The American reusable **Space Shuttle** was launched in 1981. The universe came into much sharper focus with NASA's launch of the Hubble **space-based optical telescope** in 1990. Striking developments over the next twenty-five years were a craft on a **solar orbit** (1992, USA and Europe), a **landing on an asteroid** (2001, USA), a **probe into deep** (or interstellar) **space** (USA), and the first **food** (lettuce) **grown – and eaten – in space** (2015, USA and Japan).

The Hubble space telescope in orbit, 2009

FINDING YOUR WAY

MAPS

The oldest map, found on the walls of a cave in Lascaux, France, is that of the **night sky** (*c.* 14500 BC). Maps of **earthly features** are found on Babylonian clay tablets (anywhere between 7000 and 2500 BC – scholars don't agree), and the Babylonians also etched the first **world map** (*c.* 600 BC) (Iraq). Anaximander of Miletus (*c.* 611 to *c.* 546 BC, Greece) is regarded as the first proper **mapmaker**. Another mapmaking Greek, Eratosthenes (*c.* 275 to *c.* 195 BC), who coined the word '**geography**', worked out the **circumference of the Earth** remarkably accurately and divided it into the five climactic zones we still use (two polar, two temperate, one tropical). A third Greek, Claudius Ptolemy (*c.* AD 100 to *c.* 170), gave us a revolutionary depiction of a spherical Earth using **perspective projection**, and suggested lines of **latitude** and **longitude**. (The idea of a **geographic grid** had also appeared in China *c.* 120 BC.) Mercator's famous **cylindrical projection** was presented in 1569. The first **road map** is said to be Egyptian (*c.* 1160 BC), and Rand McNally (USA) published a modern equivalent in 1904. Though **globes** had apparently been made since ancient times (e.g. *c.* 150 BC by Crates of Mallus, Turkey), the first to survive was made by Martin Behaim in 1492 (Germany). The earliest book we would recognize as an **atlas** was printed in the Netherlands in 1570.

NAVIGATION – LATITUDE AND LONGITUDE

Ever since the Mesopotamians, Persians and Greeks devised our present navigational system of **360° in a circle** and days of **hours, minutes and seconds** (*c.* 3500–500 BC), and realized they could **navigate by the stars**, travellers have discovered new and more accurate ways of ascertaining their position and direction of travel. Mediterranean **nautical charts** date back to the sixth century BC, roughly when the first **lighthouse** was built at Sigeum (Turkey). Among the first navigational devices were the **astrolabe** (*c.* 150 BC, Greece), the **quadrant** (*c.* AD 150, Greece), and the **magnetic compass** (*c.* 1100, China). The **cross-staff** was invented in the fourteenth century. With the development of **trigonometry** (second century BC–fifteenth century AD), the Portuguese were able to produce tables for determining **latitude by the angle of the noonday sun** (*c.* 1480). Various **logs** (fifteenth to sixteenth century) enabled speed to be calculated with greater precision, and with the invention of the **marine chronometer** (see p.182) **longitude** could be worked out as accurately as latitude.

NAVIGATION – ELECTRONICS

The **sextant** (1757, UK) and the establishment of the **Greenwich Meridian** (1851, UK), together with twenty-four codified **time zones** (1884, Washington, D.C., USA), further increased navigational accuracy and facility. Modern electronics then produced a revolution in navigation. Italian scientists developed a prototype **radio direction finder** (RDF) system in 1907, and the UK set up more sophisticated **radio beacons** from 1921. **Radar** was invented in 1935 (UK) and **installed on a ship** two years later (USA). Next came **satellite**

navigation, launched by the US Navy in 1964. By 1985, the first fully functional military **GPS** was up and running (USA). The Japanese led the way in automated civilian navigation for cars with built-in **GPS** (1990), **automatic parking** (1999) and **in-car voice assistance** (2017). In 2007, **Google Street View** was launched, with the photographing of the streets of several US cities.

Notable first journeys on land (and ice) include Alexander Mackenzie's expedition crossing North America (Scotland, UK) in 1792–3, the 1860-1 crossing of Australia by Burke (Irish) and Wills (UK), walking around the world (disputed between George Matthew Schilling, 1897–1904, USA, and Dave Kunst, 1970–4, USA), reaching the North Pole (also disputed between, among others, Frederick Cook, USA, and two Inuits, Aapilak and Ittukusuk in 1908, and Robert Peary, Matthew Henson (both USA) and four Inuits, Ootah, Seegloo, Egingway, and Ooqueah in 1909, reaching the South Pole (expedition led by Roald Amundsen, Norway) in 1911, and climbing Mount Everest (Edmund Hillary, New Zealand, and Tenzing Norgay, Nepal, in 1953). Five great first sea voyages deserve mention: the unconfirmed circumnavigation of Africa by Phoenician sailors (*c.* 600 BC), Admiral Zeng He's remarkable voyage from China to the Middle East and Africa (1405–7), the Viking Leif Erikson's crossing of the Atlantic via Iceland and Greenland (*c.* AD 1000), the Spanish-funded voyage from Spain to the Caribbean by the Italian captain Christopher Columbus (1492), and the Portuguese captain Ferdinand Magellan's crossing of the Pacific on a Spanish-funded expedition that eventually circumnavigated the globe (1519–21). Finally, there are the famous first flights: non-stop across the Atlantic by John Alcock and

Arthur Brown in 1919 (UK), solo from Europe to Australia, by Amy Johnson in 1930 (UK), and around the world non-stop by James Gallagher and crew of thirteen in a Boeing Superfortress (USA) in 1949.

PART V: SCIENCE AND ENGINEERING

TOOLS AND FIXINGS

TOOL USERS

If *Homo habilis* was the earliest manifestation of the human species – and this has been challenged – then we have been tool users for well over 2 million years. Not that we were wielding screwdrivers back then – the first tools were simply pieces of chipped flint (often called **hand axes**) with a sharp edge for cutting or scraping. The origins of the **spear** are even more obscure. Apparently, chimps have been making wooden spears for millions of years, suggesting that early humans did so too; nevertheless, we have no firm evidence of a human-made spear before *c.* 450,000 BC. Sophisticated **stone tips** were fitted perhaps 200,000 years ago. Axes with **ground blades** have been around for at least 44,000 years (Australia), and hafts were probably attached shortly after this – the first secure fitting is dated *c.* 6000 BC (Asia). By this time, **bows and arrows** had been in use in Germany for some four millennia. The story of the **hammer** is pretty similar to that of the axe, though it may have started a million years earlier; there is evidence of **hafted hammers** from *c.* 30,000 BC.

CARPENTRY TOOLS

Carpentry was first practised in the Near Eastern Fertile Crescent and the Indus Valley. The ancient Egyptians gave us the **saw** (*c.* 3000 BC), the **adze** (*c.* 2500 BC), the **rasp** (bronze, *c.* 1200 BC), and probably the **spokeshave** (date unknown). They also created **veneer** about 5,000 years ago. **Drills** were used in Mehrgarh (*c.* 3500 BC, Pakistan) and stone **chisels** in Neolithic Germany (*c.* 3500 BC). The Romans fitted a **wooden handle onto the saw** and are likely to have invented the **plane**. We have images of metal **pliers** from ancient Greece (first millennium BC).

FIXINGS

The first fixings, going back tens if not hundreds of thousands of years, were **bindings** of creeper, leather or pliable twigs. **Mortice and tenon joints** – sometimes held by pegs – were employed in German, Egyptian and Chinese woodwork *c.* 5000 BC, and in the construction of Stonehenge *c.* 2500 BC. The ancient Egyptians fashioned the metal **nail** about 3400 BC and made **animal glue** some 2,000 years later. **Bolts** for securing a door go back to Roman times (see p.8), as possibly does the idea of threading the bolt to make a **screw that was** tightened with a spanner. This was certainly around in fifteenth-century Europe but the **screwdriver** didn't appear until 1744. Frenchman Jacques Besson (*c.* 1540–73) built a **lathe for making screws and bolts**, though the process was not properly mechanized until Jesse Ramsden's lathe of 1770

Early mortice and tenon joints were employed in the construction of Stonehenge, seen here during restoration, *c.* 1920

(UK) and David Wilkinson's mass production techniques of 1798 (USA). Though the Canadian Peter Robertson invented the **square-headed** (cross-headed) **screw** in 1908, it is more commonly associated with the American businessman Henry Phillips. Another US businessman, William G. Allen, patented the **hexagonal Allen key** in 1909 or 1910.

MACHINERY

MECHANICAL ASSISTANTS

The first **clothes iron** – a metal pan filled with hot coals (first century BC, China) – was hardly high-tech. Nor were the slabs of heated iron (**sad irons**) used in Europe from the seventeenth century onwards. During the nineteenth century, the **self-heating** iron appeared in the USA and elsewhere, fired by kerosene, natural gas or even (dangerously!) petrol. The **electric iron** was made in the USA in 1882, and a **thermostat** added *c.* 1926. **Electric steam irons** went on sale in 1926, with **cordless** versions in 1984 (both USA). Hand-powered **sewing machines** are talked about in the panel to the right; the **electric sewing machine** went on sale in 1889 (Singer, USA). In 1589, the English cleric William Lee invented the **knitting machine** (or stocking frame), and around 1806 the Frenchman Pierre Jeandeau pioneered future improvements with the concept of a **latch needle**.

TYPEWRITERS

Identifying the first typewriter is not easy: Henry Mill (1714, UK), Pellegrino Turri (*c.* 1808, Italy; Turri had also invented carbon paper in 1801), William Austin Burt (1829, USA), Father Francisco João de Azevedo (1861, Brazil), and Peter Mitterhofer (1864–7, Austria) all have a claim. Thereafter, developments came thick and fast. The year 1872 saw the invention of the electric typewriter by Thomas Edison (USA), with the QWERTY keyboard (USA) arriving in 1874, and the 'golf ball' typewriter (IBM Selectric, USA) in 1961. Electronic machines appeared in the 1970s (Diablo, USA), shortly before the clattering world of the typewriter was swept away by the computer–printer revolution (see p.239).

SEWING SQUABBLES

As we all wear clothes at some time or another, and each article needs to be stitched together, there was always a fortune to be made from the invention of a sewing machine. Englishman Thomas Saint patented one in 1790, but the idea failed to take off. Forty years later, the French tailor Barthélemy Thimonnier devised a machine for chain stitching. John Fisher's machine of 1844 (UK) went one better by using two sources of thread. However, he bungled his patent application and the following year his lockstitch design (patented 1846) was stolen by the American Elias Howe. In 1851, the design was pinched and patented once more, this time by fellow American Isaac Merritt Singer. Howe sued – and won – forcing Singer to share his profits. As a result, when Singer's firm grew into a household name, its founder, and the man whose idea he had stolen, both became millionaires.

POWER TOOLS

Though the invention of the **circular saw** is said to have been inspired by the quizzical ponderings of Tabitha Babbitt in 1813 (USA), the modern era of power tools really begins eighty-two years later with the **electric drill** (C. & E. Fein, Germany). Weighing 7.5 kg, it was not exactly DIY kit! The next significant move came when Duncan Black sold his car and teamed up with his friend Alonzo Decker to manufacture **pistol-grip, trigger-switch electric drills** (1916, Black & Decker, USA). Many American gadgets and accessories followed, including the same firm's **electric screwdriver** (1923) and **cordless drill** (1961). Circular power drove **floor sanders** (1916), and took a deadlier form in the **chainsaw** 'tree-felling machine' (1936, both USA). The American George McGill made a **stapler** in 1866, paving the way for the **staple gun** (1934, USA). Morris Pynoos came up with the more powerful (and deadly) **nail gun** (*c.* 1944, USA). **Steam pressure washers** first squirted in 1926 and a **domestic pressure washer** entered the shops in 1950 (both USA). Finally, ensuring the US did not have a total monopoly of the gadgetry market, the Spanish Boada brothers devised a **tile cutter** in 1951.

METAL AND MACHINERY

SMELTING

We had to get hold of metal before we could start using it. The first was **gold**, which – together with occasional amounts of copper – could be gathered in its 'natural' state. The other

first metals (tin, lead, silver, iron and mercury) were obtained by applying heat to their ore. This **smelting** process is thought to have begun with **tin** and **lead** (*c.* 6500 BC, Turkey). Then came copper and iron (see p.18), leading to the production of **steel** (*c.* 1800 BC, Anatolia/Turkey).

BLASTING AND JOINING

Iron and steel manufacture was greatly improved by the **blast furnace** (fifth century BC, China), which underwent further development when **coke** replaced charcoal (1709, UK). **Puddling** iron to make a more malleable product was first done in China about 2,000 years ago, but a proper **puddling furnace** had to wait until 1784 (UK). Other significant firsts in steel manufacture were the production of **crucible steel** (eighth century BC, India and Central Asia), the **Huntsman process** (1740, UK), the **Bessemer Converter** (1858, UK), the basic oxygen process (1948, Austria) and the **Corex process** (1970s, Austria); and in iron production, the modern **rolling mill** (1783, UK) and **extrusion** (1797, UK). **Galvanizing** began in France in 1836; **stainless steel** was invented in the UK in 1913, and **soldering** dates back to the discovery of tin some 4,000 years ago, with the **electric soldering iron** appearing in 1921 (Germany). **Welding** is said to have started in ancient Greece in the fifth century BC, with **electric arc welding** invented by a Russian and a Pole in 1881–2 (Russia–Poland), and **oxygen-acetylene welding** in 1903 (France). The creation of 3D metallic components by **laser cladding** was first done in 1978 (USA).

METALS

Many metals were known in compound form long before they were identified. The first to be distinguished specifically was copper (*c.* 8700 BC), followed by lead (*c.* 6500 BC), iron (*c.* 5000 BC), silver (*c.* 4000 BC), tin (*c.* 3500 BC), gold (*c.* 3000 BC), and mercury (*c.* 2000 BC). More recently, we identified chromium (1780), uranium (1789), silicon (1824), aluminium (1824–5), radium (1898), and americium and Curium, the first elements that do not occur naturally on Earth (1944).

MILLS AND FACTORIES

MILLS AND MACHINERY

Bread being the staff of life, it is not surprising that the process of bread making gave rise to simple **machinery**: the **rotary grinding mill** powered by an ox, donkey or horse (fourth century BC, Carthage, Tunisia). After a brief period – or possibly simultaneously – humans and beasts began turning **irrigation water wheels** in India and the Middle East. By the third century BC, in the Near East the process had been reversed to produce **overshot water wheels** for milling. The earliest record of a **gear mechanism** comes from China in the fourth century BC. **Undershot waterwheels** were certainly turning in Hellenistic times (*c.* 320 BC onwards). Most agree that the idea of a **windmill** originated in Persia in the first millennium AD, though it is possible that the Babylonian emperor Hammurabi (*c.* 1810 to *c.* 1750 BC) employed them for irrigation. **Masonry mills**, with just the cap and sails

The first offshore wind farm in Vindeby, Denmark

rotating to face the wind, were constructed in thirteenth-century Europe. An **electricity-generating wind turbine** started spinning in Scotland in 1887; an **automatic** version appeared the following year in Ohio, USA. Denmark opened an **offshore wind farm** in 1991.

FACTORIES

When is a factory not a factory? If we rule out lots of slaves working together (as happened in ancient Rome, for example) and include **mass-production**, an **assembly line** and **manufactured components**, then a good claim for the first **factory** can be made for the Venetian Arsenal, founded in 1104 (Italy), in which 16,000 workers turned out almost a ship a day. On the other hand, if we take a more modern definition of many workers in a single building using machinery to mass produce, the first factory was either Lombe's **silk throwing mill** in Derbyshire, UK (1718–21) or Richard Arkwright's Cromford Mill, a **cotton spinning factory** also in Derbyshire, UK (1771). The Soho Manufactory in Birmingham, UK, was probably the first driven by **steam power** (1782); Étienne Lenoir's gas-fired **internal combustion engine** of 1860 was widely sold for industrial use (Belgium), and **electric motors** are recorded as powering factories in 1889.

ASSEMBLY LINES

The British Royal Navy's Portsmouth Block Mills operated the first **continuous, linear assembly plant** (1795–8), while the first **completely automated manufacturing** process was an American flour factory (1916). In the meantime, the **assembly line** concept, born in an English steam engine factory in 1853, had crossed the Atlantic to be used for **mass disassembly** in a Chicago slaughterhouse (1867, USA), and upgraded with the employment of an **electric conveyor belt** in an American canning factory (1885). Ransom Olds built an **assembly line for motor cars** (1901, USA), and twelve years later Henry Ford's **moving assembly line** was churning out Model Ts quicker than the time it took for the car's (black) paint to dry. An **industrial robot** was created as early as 1938 (USA), but the version installed by General Motors in 1961 was the first to be employed on a **large scale**.

MACHINE TOOLS

The first and simplest machine tool was probably the **bow drill**, found in Mehrgarh (Pakistan, see p.136) around

6000 BC. The **potter's wheel** (see p.17) came next, followed by the ancient Egyptian **lathe** (*c.* 1300 BC). **Pole lathes**, operated by a treadle, were found in ninth-century Europe. The great leap forward in machine tooling came with the Industrial Revolution, starting in the mid-eighteenth century. The innovative Frenchman Jacques de Vaucanson made a **rotary file** and **all-metal lathe** (*c.* 1760). The modern **screw-cutting lathe** followed in 1775 (UK), and the **metal-working lathe** emerged out of the horse-powered cannon boring machine of 1772 (UK). Another Englishman, Joseph Bramah, invented the **hydraulic press** in 1795, and in the same year the American clockmaker Eli Terry devised the modern **milling machine. Metal planers** were made in early nineteenth-century England, and the **cylindrical grinder**, essential for the development of all forms of engine, was made in the US in the 1830s.

LOGISTICS, MANAGEMENT AND SUPPLY

The first **logisticians** were the officers responsible for supplying the armies of ancient times with men, food and weapons. The Romans called them *logistikas* – hence the **word** 'logistics', first appearing in French (*logistique*, 1830 or earlier) then English (1846). People began talking of the **supply chain** in 1910 and **supply chain management** in 1982. By then, the term **planned obsolescence** had been around for fifty years, and practised in the bicycle and car industries for even longer. Other logistics concepts appeared as the process became larger and more complex: **postponement** (1950), **material requirements planning** and the **bullwhip effect** (both 1961), **reverse logistics** (1992), and **continuous replenishment** (1997). To help people navigate through this increasingly pressurized and jargon-filled world, the first **business school** opened in the US in 1881, and the first **postgraduate qualification** (a Master of Science in Commerce) was awarded in 1900. Eight years on, the Harvard Graduate School of Business Administration launched the **MBA**. Frederick Winslow Taylor's *The Principles of Scientific Management* (1911) was the first major work on the **theory of modern management**, while logistics was recognized as a **profession** in 2000. Back on the factory floor, the **lifting truck** was at work in 1887, a **self-powered** version in 1906 (both USA), and the first capable of **horizontal and vertical movement** in 1915 (UK). By the end of the 1920s, the **forklift** and the **pallet** were commonplace. **Warehouses** first served the cities of the ancient world; **automated storage and retrieval systems** were installed two millennia later (1960s). The **barcode** (see p.33) and the standard **container** (see p.112) completed the picture.

ENGINES

PUMPING AND HAULING

The English word 'engine' appeared in the fifteenth century, to mean a **mechanical contrivance** of some sort. Its unspecific embrace might include the water-lifting **shaduf** (3100 BC, ancient Egypt), the **pulley** (*c.* 1500 BC, origin unknown, possibly Mesopotamia), the **crane** (*c.* 500 BC, ancient Greece), the **winch** or **capstan** (*c.* 500 BC, ancient Greece or Assyria – Iraq/Syria), the **block and tackle** (attributed to the Greek genius Archimedes), any kind of **mill** (see p.142), the **screw pump** or **suction** or **piston pump** (*c.* 275 BC, probably ancient Greece), the **treadwheel** (first century AD, Rome), the **windlass** (attributed to Archimedes but first depiction AD 1313, China), and **gearing** (see p.142). In 1654, the German Otto von Guericke invented the **vacuum pump**; a **diaphragm pump** was patented in 1854, and a **rotary vane pump** twenty years later (both USA).

STEAM

Heron of Alexandria's whirling aeolipile of the first century AD was, apparently, the first **device powered by steam**. It didn't do anything useful, unlike the rudimentary **steam turbine** constructed by the Ottoman Taqi al-Din to turn a spit (*c.* 1551). A working **steam pump** was made by the Spaniard Jerónimo de Ayanz y Beaumont in 1606. Thomas Savery was responsible for the first **effective steam pump**, 'The Miner's Friend', powered by steam and a vacuum (1698). Frenchman Denis Papin (1647 to *c.* 1713) devised the **safety**

A steam turbine at Carville Power Station, 1907, UK

valve; starting in 1712, the Englishman Thomas Newcomen constructed the first **commercially successful steam pumps**. From 1776, James Watt and his business partner Matthew Boulton were producing the first **true steam engines** capable of powering machinery (UK). John 'Iron-Mad' Wilkinson's **boring machine** (1774, UK) produced high-quality cylinders. The Watt-Boulton team added the **sun and planet gear** for turning a piston's up-and-down movement into circular motion (1781), the **governor** (1788), and, with Cornishman Richard Trevithick (see p.107), the **double-action piston**. Trevithick also gave us the **high-pressure steam engine** (1800), while Jean-Jacques Meyer developed the **expansion valve** (1841, France) and Sir Charles Parsons a **modern steam turbine** capable of generating electricity and powering a ship (1884, UK).

CRANKS AND COMBUSTION

The Dutch genius Christiaan Huygens (1629–1695) came up with the idea of an **internal combustion engine**, while Nicéphore and Claude Niépce (1807, see opposite page) and the Swiss engineer François Isaac de Rivaz, whose hydrogen-powered engine was fitted to a vehicle (1807), pioneered working machines. Samuel Brown's **gas vacuum engine** was the first to serve industry (1823, UK); the American Samuel Morey invented the **carburettor** for his engine (1826); William Barnett came up with the idea of **in-cylinder compression** in 1838 (UK); and a **twin-cylinder** industrial engine was manufactured in Italy in 1856. Having made their debut in Han, China (*c.* 202 BC to AD 220), **cranks** were joined to a **connecting rod** in Turkey (third century AD), where they were incorporated into a **crankshaft** in 1206.

THE PYRÉOLOPHORE

The French Niépce brothers, Nicéphore and Claude, were nothing if not inventive. In 1807, they installed what is claimed to be the first internal combustion engine – the Pyréolophore – in a boat on the River Seine. It worked by a series of mini explosions (approximately one every ten seconds) created by a mixture of moss, coal dust and resin inside a copper chamber. The power generated by the combustion was used to draw river water in through the front of the boat and force it out through a pipe at the back, driving the vessel along. The engine, though ingenious, was not a success. When the patent granted by Napoleon I expired, Claude went to London, received a new patent from King George III, and continued to push his doomed Pyréolophore until he went mad and died in 1828. Nicéphore had remained in France where he focused his creative mind on photography (see p.173).

THE INTERNAL COMBUSTION ENGINE

The engine built by Jean Joseph Étienne Lenoir (see p.107) was virtually a steam engine running on gas. The Austrian Siegfried Marcus developed the mobile **gasoline engine** in 1870, with a **four-stroke** machine following six years later (see p.108). The **two-stroke** engine with in-cylinder compression was patented in 1881 (UK), and *c.* 1884 the

British engineer Edward Butler came up with a string of firsts: the **gasoline-powered** internal combustion engine, the spark plug, ignition magneto, coil ignition, spray jet carburettor, and the word '**petrol**'. Benz's **motor car** (see p.108) appeared in the same year as Gottlieb Daimler's **supercharger** (1885, Germany). After Rudolf Diesel's **compression ignition engine** (patented 1892, Germany), what followed were largely refinements of the original concept until Felix Wankel came up with the **rotary engine** in 1954.

NUCLEAR POWER

Ernest Rutherford was the first scientist to realize that vast amounts of energy were released when lithium **atoms were split** (1932, UK). Following the discovery of the **neutron** in the same year and laboratory, the Hungarian Leó Szilárd understood the possibility of a **chain reaction** (1933). Building on this pioneer work, the first **nuclear reactor**, Chicago Pile-1 (USA), became operational in 1942; the first **nuclear weapon** was **tested** in July 1945; and a **nuclear weapon used in anger** on the Japanese city of Hiroshima the following month. The first test of a **hydrogen bomb** took place in 1952 (USA). **Electricity** had been **generated** by nuclear power the previous year, and **nuclear-powered ships** soon followed (see p.123). **Large-scale generation of electricity** by nuclear power was pioneered in Russia (1954) and was operating on a **commercial basis** in the UK in 1956. It was not long before there were accidents. The first **major catastrophe**, and one of the most devastating, was the Kyshtym Disaster of 1957 – hushed up by the Soviet authorities for thirty years.

ELECTRICITY

EARLY SPARKS

Linguistics provides a possible clue to the first realization that **lightning** – the most vivid natural display of electrical power – is **electricity**: in the late-medieval Arab world the word for the electric ray (fish) and lightning were the same. After the **coining of the words** 'electric' and 'electricity' in 1642, linking the phenomenon to amber ('*electrum*' in Latin), Benjamin Franklin (1706–90, USA) confirmed Arab suspicions about the nature of lightning, and suggested electricity had **positive** and **negative** manifestations. In 1780, the Italian Luigi Galvani discovered the presence of **electricity in animals' bodies**; another Italian, Alessandro Volta, made the first **battery** in 1800; and twenty years later, Hans Christian Ørsted and André-Marie Ampère (Denmark, France) recognized **electromagnetism**. The **electrical circuit** was defined in terms of current and resistance by Georg Ohm (1827, Germany), and in 1831 Michael Faraday worked out the principle of **electromagnetic generation** (UK).

GENERATION

The Frenchman Hippolyte Pixii built the first **alternator** and **commutator** in 1832, and an **industrial generator** went into operation in 1844 (UK). Antonio Pacinotti, an Italian professor, enabled smooth generation of electricity with his **ring armature** (1860), while the Hungarian Ányos Jedlik's proposal for **dynamo self-excitation** (1856) paved the way for the **practical dynamos**, invented independently by Samuel Varley (UK), Sir Charles Wheatstone (UK), and Werner von Siemens (Germany). **Hydroelectric power** was first generated in 1870 (Cragside, UK), and the first **power station** providing electricity for the public began operating in Godalming (UK) in 1882, the same year as **coal-fired** and **steam-powered stations** opened (UK and USA). **Alternating current** was first used practically in 1855 (France), and an **AC power station** was constructed in Britain in 1866. In 1953, electricity was generated by a **silicon solar cell** in the laboratory, and commercially in 1956. **Tidal power** had driven water mills (see p.142) for many centuries, but a **large-scale tidal power plant** did not open until 1966 (estuary of the River Rance, France). Italy pioneered **geothermal power** in 1904 and opened a **geothermal power station** in 1911.

ELECTRIC POWER

The first (very simple) **electric motor** is attributed to Andrew Gordon, a Scottish monk living in Bavaria (*c.* 1745). After the principle of **mechanical force arising from the interaction between electricity and a magnetic field** had been described by André-Marie Ampère (France) in 1820, it was demonstrated by Michael Faraday (UK) the following year. The Hungarian physicist Ányos Jedlik produced the first **true electric motor** in 1828, and William Sturgeon (UK) the first **motor able to turn machinery** in 1832. The invention of the **ring armature** (1864, Italy) led to the electric motor being widely adopted as a source of stationary and mobile power. **Solenoids** date from the 1820s (André-Marie Ampère, France) and **transformers** from 1836 (Ireland). Michael Faraday (UK) first noted the **semiconductor** effect in 1833, and Jagdish Chandra Bose (India) produced the first **solid-state device** in 1901.

MATERIALS

WOOD, STONE, CLAY

The first materials used by human beings were those that came to hand most readily: **wood and stone**. **Clay** came next, the oldest examples of **ceramic work** being Gravettian female figurines of around 29,000 BC (France). Then came basic clay pots (see p.17), **stoneware** from the Indus Valley civilization (*c.* 2500 BC, Pakistan), and porcelain (see p.17). Bricks of **dried mud** (*c.* 7500 BC) first appeared either in Anatolia (Turkey) or the Indus Valley (Pakistan). The Chinese were the first to **fire bricks** (*c.* 4400 BC). From about 1400 BC, natural **rubber** balls were bouncing around in a Meso-American ball game, and, as far as we know, the people of Malaysia were making things like knife handles out of **gutta-percha latex** (used in the rest of the world from the 1840s). **Metals** were probably smelted around 8,500 years ago (see p.140). **Plaster** was employed in Jordan *c.* 7500 BC, and **stucco** decoration shortly after this. **Cement**'s origins are obscure, but there is evidence of its use in Crete some 4,500 years ago. **Concrete**'s early history is similarly obscure, though we know the Greeks were building with it *c.* 1300 BC.

THE PLASTIC REVOLUTION

As artificial fabrics are covered elsewhere (see p.37), we now turn to other materials human beings have created. The **vulcanization** process for hardening rubber was patented by Thomas Hancock (UK) in 1843, a few weeks before Charles Goodyear did the same in the US. Using plant cellulose,

A Bakelite telephone, 1931

Alexander Parkes (UK) manufactured the first **new solid material** ('Parkesine', later 'celluloid') in 1862, and in 1872 a patent was taken out for a plastic injection moulding machine (USA). The first wholly artificial material appeared in 1907, when the Belgian-born American Leo Baekeland used fossil fuels to make a synthetic plastic, which he named **Bakelite**. This was followed by numerous other materials, most of which had been discovered in the nineteenth century but never exploited commercially: **PVC** (Polyvinyl chloride, or just 'vinyl', discovered in Germany in 1872 and manufactured in the USA in 1926), **polystyrene** (discovered in Germany in 1839, manufactured there in 1931, and in the USA in its expanded form in 1954), **polythene** (polyethylene, first synthesized in Germany in 1898 and manufactured in the UK in 1933), **polypropylene** (discovered in the USA in 1951 and manufactured in Italy in 1957), and **melamine** (invented in Germany *c.* 1835 and manufactured in the USA as the **laminate** 'Formica' in 1913).

A PLASTIC WORLD

Plastic led to the creation of a host of familiar products: **Scotch tape** (1930, USA), **fibreglass** (1942 or earlier, USA and Germany), **superglue** (1942, USA), **acrylic paints** (1940s, Germany), **expanding polyurethane insulation foam** (Germany, 1940s), **Tupperware** (1946, USA), **Polly-filla** (1959, UK), **silicon breast implants** (1961–2, USA), **polythene shopping bags** (1965, Sweden), and the **Swatch** (1983, Switzerland). More recent developments in the materials field have included the creation of the first synthetic nanomaterial, the **carbon fullerene**, in 1985 (UK and USA), and the production of **graphene**, a single-atom-thick crystallite of graphite (2004, Russian scientists working in Manchester, UK). ICI launched commercially available **biodegradable plastic**, Biopol, in 1990 (UK); fifteen years later Coca-Cola launched the first **plastic bottle made from plant materials**.

LIGHT

SEEING AND BELIEVING

The Greeks and Chinese were the first to study light objectively. The Chinese philosopher Mozi is credited with mentioning the **camera obscura** (projector or pinhole camera) idea *c.* 400 BC; in the next century, Euclid – more scientifically – studied the mathematics of **reflection** and suggested light **travelled in straight lines**. Over a millennium later, the Dutchman Christiaan Huygens (1629–95) posited the theory of light as a **wave** (proved *c.* 1800 by the Englishman Thomas Young), and *c.* 1672 Sir Isaac Newton was the first to demonstrate that white light was a **mixture of colours**. The idea of something from which no light could escape, now known as a **black hole**, originated with the French Marquis de Laplace (1749–1827), while another French scientist, Léon Foucault, was the first to measure the **speed of light** accurately (1850). In the next decade, James Clerk Maxwell (UK) first linked light to **electromagnetism**.

In 1905, the German–Jewish genius Albert Einstein turned most previous theories on their head when he said that light exists as **photons** (particles) *and* – according to his theory of **special relativity** – a **field of waves**. By this time, he had left most non-scientists struggling in his wake.

CANDLES AND OIL

Fire (see p.13) was obviously the first source of artificial light. No one knows precisely when or where fire was trained to create a crude **lamp** – fatty moss burning in a shell or rocky hollow? – but it may have been around 70,000 years ago. Sixty thousand years after this, we get the earliest purpose-made **stone lamps** (France), and by 4000 BC we have **pottery lamps** (Europe and Near East). The Egyptians began making **candles** from beeswax about 5,000 years ago, and candles from fats and other waxes followed. Al-Rāzi (see p.61) described a lamp burning **mineral oil** in the ninth century AD, but the idea was of little use before the discovery of **kerosene** in 1846 (Canada) and the sale of the kerosene lamps seven years later (Poland and USA). Other significant firsts were the **miner's safety lamp** (1815, UK), **pressure (Tilley) lamps** (1818, UK), **limelight** (1820s, UK) and the (gas) **mantle** (1881, France).

SWITCHING ON

Gas lights first flared in 1792, and within a decade they had been joined by the **electric arc light** (both UK). The **long-lasting incandescent electric light bulb** created by Thomas Edison (USA) and Joseph Swan (UK) in 1879 was the result of many other developments, including James Lindsay's demonstration of an **electric light bulb** bright enough to read by (*c.* 1835, UK), Marcellin Jobard's **vacuum bulb** (1838, Belgium) and Alexander Lodygin's **nitrogen-filled bulb** (1872, Russia). The Edison-Swan bulb was improved with a **tungsten filament** (1904, Germany and Croatia) and **coil filament** (1917, USA). Philips manufactured the **energy-saving bulb** in 1981 (Netherlands). Meanwhile, a range of other forms of electric light had sprung up: a primitive **fluorescent lamp** (1867, France), the **gas-discharge lamp** (1894, preparing the way for neon lighting and plasma-screens, USA), the **mercury vapour lamp** (1901, USA),

A nineteenth-century Swan-Edison light bulb

the **sodium vapour lamp** (1920, USA), the **light-emitting diode** (LED, 1927, Russia), the **halogen light** (1953, USA) and the **laser** (1960, USA). The Englishman David Misell made a battery-powered electric torch in 1899. Gaston Plante invented the **lead-acid rechargeable battery** in 1859 (France), and the modern rechargeable **lithium-ion battery** was created in 1985 (USA).

FAR AWAY AND UP CLOSE

The Dutchman Hans Lippershey built the first **telescope** in 1608, and Galileo Galilei (Italy) made the first **astronomical telescope** the following year. Isaac Newton's **reflector telescope** is dated 1668 (UK). Colour distortion was reduced with the **achromatic lens** (1733, UK), and much greater telescopic power was possible with an effective **Cassegrain reflector** (*c.* 1910, USA and France), based on an idea developed by the French priest Laurent Cassegrain (*c.* 1629–93). **Radio telescopes** began with Karl Guthe Jansky (1932, USA). At the other end of the scale, **magnifying lenses** were known to the ancient Greeks (fifth century BC) and described by the Arab polymath Hasan Ibn al-Haytham in AD 1021 (Iraq). Dutch spectacle makers built the earliest **compound microscopes** around 1620, although a number of individuals may have had the idea before this. The Italian Giovanni Faber was the first to speak of a '**microscope**' when he employed the word in 1625 to describe an early device made by Galileo. The first really effective microscope (300 x) was built by Antonie van Leeuwenhoek (1670s, Netherlands – see p.62). The **electron microscope** was built in 1931, with a **scanning version** in 1935 (both Germany).

BUILDINGS

EARLY CONSTRUCTIONS

The first **houses**, as noted, were built around 11,000 years ago (see p.8). It is now thought that the earliest **religious building** – a kind of **temple** in Gobekli Tepe, Turkey – was constructed at about the same time. Just a thousand years later (8000 BC), the city of Jericho (Israel/Palestine) was surrounded by the first **defensive walls**. **Stone castles** – permanent defensive positions for individuals or families – probably went up not long after this, but the Citadel of Aleppo is the oldest known (*c.* 3000 BC, Syria). The first **palace** – at Thebes in Egypt – had been constructed a century before this. The Romans built large homes known as **villas** (second century BC, Italy), and another Roman innovation from this time, the **mansion** (the English word first appears in the fourteenth century), morphed into a 'palace' when situated on Rome's Palatine Hill. Rome's great rival Carthage may have been the first city to build **apartment blocks** several storeys high (*c.* 300 BC, Tunisia). The first purpose-built **church** is said to be that dedicated to St George in Rihab, Jordan (*c.* AD 230), while the first purpose-built **mosque** was either Al-Sahaba (*c.* AD 613, Eritrea) or the Quba in Saudi Arabia (*c.* AD 622).

MATERIALS AND MOVERS

Traditional building materials – **wood**, **stone**, **brick**, **plaster**, **cement** and **concrete** – are dealt with elsewhere (see p.153). **Cast iron** was used for pagoda-building in early medieval China (seventh to tenth centuries BC); Ditherington Flax Mill (1796, UK) was the world's first **iron-framed building**, and the Commissioner's House of the Royal Naval Dockyard, Bermuda

America's first skyscraper, the Chicago Home Insurance Building, 1885, USA

(1820s) was the first **house with an iron frame**. Though the Romans had grown vegetables under frames, the earliest proper (i.e. heated) **greenhouses** were found in Korea during the 1450s. The Crystal Palace, built for the Great Exhibition of 1851 (UK), paved the way for the extensive deployment of **plate glass within metal frames**. Across the Channel, the Frenchman François Coignet erected the first reinforced concrete structure (a four-storey house) in 1852, and Chicago's Home Insurance Building, the first **skyscraper**, was also the first building entirely supported by its **steel skeleton** (1885, USA). Despite reports of the Greek scientist Archimedes (third century BC) building a **lift** (or **elevator**), and excluding a wide variety of early modern raising devices, the first proper lift was either London's steam-powered 'ascending room' of 1823 or a belt-driven, counterweight machine of 1835 (both UK). **Safety lifts** were devised in Italy (1845) and the US (1852), while in 1880 the German engineer Werner von Siemens built the **electric lift**. Thirteen years later, Chicago installed a **moving pavement** or **walkway** (USA).

CROSSINGS AND TUNNELS

Early **bridges** were, presumably, logs or tree trunks across streams or small rivers. The first known **constructed wooden bridge** (as opposed to stepping stones or marshy walkways) was that across Lake Zurich (1523 BC, Switzerland) and the first **stone bridge** was probably built in Greece *c.* 1300 BC. Even earlier (*c.* 2000 BC), the Minoeans of Crete had built **aqueducts**. The first **iron bridge** was built in Shropshire, England, in 1779 (UK). According to ancient Chinese sources, the earliest **rope suspension bridges** were across the rivers and ravines of the Western Himalayas (first to second centuries AD, Afghanistan, India, Pakistan, Tibet); the first **iron-chain suspension bridge** was made in Tibet in 1433. The US originated the **wire suspension bridge** (1847) and **steel-wire suspension bridge** (1883, Brooklyn Bridge, New York). **Concrete bridges** were a French invention (1850s).

The iron Gaunless Bridge on the Stockton and Darlington line (1823, UK) was likely to have been the first **railway bridge**, and the Hassfurt Bridge over Germany's River Main was the first modern **cantilever bridge** (1867). The idea of a **moveable bridge** goes back at least to the **drawbridges** of medieval Europe (*c.* AD 1000); over the years, designs improved with counterbalancing weights, but the massive **swing bridge** that crossed the River Tyne (UK) in 1876 was something entirely new. Hubert Gautier (France) wrote the first **book on bridge engineering** in 1716. The first **canal tunnel** was dug in France in 1679, and the first **railway tunnel** in England (UK) in 1793. (For **Road tunnels**, see p.117.)

ARCHITECTURE

According to one's definition, the world's first **architect** was either Imhotep, the ancient Egyptian designer of the Step Pyramid (*c.* 2667–2648 BC), or the Roman engineer and builder Vitruvius Pollio (*c.* 80 to *c.* 15 BC, Italy). The **word architect** first appeared in English in 1563, and the idea of **architecture as a specific skill** was first promoted by the Frenchman Philibert Delorme in 1567. Building on his initiative, the world's first **school of architecture** – the École des Beaux-Arts – was founded in the French capital of Paris in 1671. RIBA, the Royal Institute of British Architects, and the occupation's first **professional body**, was established in London in 1834 (UK). Pioneering **female architects** were Katherine Briçonnet (*c.* 1494–1526, France) and the American Julia Morgan (1872–1957), the latter being the first **female to graduate** from the École des Beaux-Arts.

ON THE LAND

DOMESTIC CROPS

The first plants to be domesticated – systematically sown and harvested (*c.* 9500 BC, Middle East) – were three cereals (barley and two types of **wheat**), four pulses (**lentil**, **pea**, **chickpea** and **bitter vetch**) and one fibre crop (**flax**). The **potato** (South America) and the **banana** (New Guinea) may have been domesticated as long ago as 8000 BC. **Sugar cane** (New Guinea) and **beans** (Thailand) date from about a millennium later, followed by **maize** *c.* 6700 BC (Mexico), **rice** *c.* 6200 BC (China) and **cotton** *c.* 3600 BC (Peru). The earliest cultivation of spices – **turmeric**, **cardamom**, **pepper** and **mustard** – appears to have been in the Indus Valley (*c.* 3500 BC, Pakistan). In all likelihood, the Chinese were the first to domesticate citrus fruits – the **citron** (lemon-like) and the **mandarin** (*c.* 2000 BC) – and they also crossed the mandarin with the grapefruit-like pomelo to produce the sweet **orange** (*c.* 350 BC, China). **Beekeeping** had begun in the Middle East about 7000 BC. **Genetic engineering** was first carried out successfully in 1972, and the first **genetically engineered food** – a tomato – appeared in 1994 (both USA).

FARMING PRACTICE

Once **agriculture** had been established (see p.7), **rotation of crops** was first carried out *c.* 6000 BC and **irrigation** around the same time (both Middle East). The **selective breeding** of animals had been going on since the domestication of the dog (see p.20), but was not done scientifically before the pioneering work of Robert Bakewell (1725–95, UK). The first **pesticide** was **sulphur** (*c.* 2500 BC, Iraq), and the first modern one was **DDT** (created 1874, Germany; discovered to be an insecticide, 1939, Switzerland). The use of animal manure as a **fertilizer** dates back some 8,000 years (Middle East). Shipments of the **mineral fertilizer** Chilean saltpetre (sodium nitrate) were made in the 1820s (Chile and UK); Sir John Bennet Lawes (UK) made the first **artificial fertilizer** in 1842, and following the synthesis of **ammonia** in 1910, the large-scale production of **nitrogen fertilizers** began within three years (both Germany).

FARMING TOOLS AND MACHINES

The first agricultural implement was some sort of tool to turn the soil (a **digging stick**, **mattock**, **hoe** or **spade**) and a **sickle** for harvesting (*c.* 9500 BC, Middle East). Harnessing oxen led to the **ard** (wooden scratch plough) *c.* 4500 BC (Iraq and Pakistan), while the **iron plough** with a **mould board** (for turning the soil) was a Chinese invention of about 500 BC. At about the same time **scythes** and **pitchforks** may have originated in Greece *c.* 500 BC and the **harrow** in China. Early images of **crooks** and **flails** – agricultural implements that symbolized authority – date from ancient Egypt *c.* 3500 BC. It is said that the Chinese made a rudimentary **seed drill** *c.* 100 BC, but the true mechanization of agriculture did not begin until the arrival of Jethro Tull's seed drill (*c.* 1701, UK), the **threshing machine** (1794, UK), the **reaper** (1826, UK), the **binder** (1872, USA), the **muck spreader** (1875, USA) and the **self-propelled combine harvester** (1911, USA).

THE COMBINE OF KALAMAZOO COUNTY

The mechanization of harvesting began with the threshing machine and the reaper (see opposite page). Around 1830, Hiram Moore (1817–1902), one of the founders of the small town of Climax in Michigan's Kalamazoo County, had the bright idea of combining the two machines in one: a single contraption that cut, threshed and winnowed its way across a field of corn. By 1835 it was ready. Amazingly, it worked, harvesting up to twenty acres a day and producing grain more efficiently than by separate reaping and threshing. So why didn't every farmer rush out and buy one? Partly because Moore did not have the capital to manufacture and market his combine harvester, and partly because it was enormous, unwieldy and hugely expensive: it took a team of twenty stout horses to haul it around the field. The first commercially successful combine – the Sunshine Harvester (Australia) – became available fifty years later.

TRACTORS

The first portable (but not self-propelled) **agricultural steam engine** was manufactured in 1839, and the **traction engine** followed in 1859 (both UK). A **petrol-engine tractor** was made in 1892, a **tracked steam tractor** in 1904, and a **petrol** version in 1906 (all USA). The first genuinely **mass-produced tractor** was the Fordson (1917, USA). In 1930,

Massey-Harris (Canada) produced a successful **four-wheel drive tractor**, and in 1933 the Irishman Harry Ferguson came up with an effective way of attaching implements to the back of a tractor using **hydraulics** and the **three arms system** still used today.

IN THE GARDEN

The first **lawn mower** was made in 1830; a **steam-driven** variant appeared in 1893, a **petrol engine** version in 1902, and a **sit-on** model around 1919 (all UK). An **electric mower** was built as early as 1926 (UK) but the idea did not catch on until after World War II; the same applied to **rotary mowers** which were pioneered in 1929 (USA) but became commercially successful only in 1952 (Australia). The **hover mower**, first produced in 1965 (Flymo, UK), was followed in 1995 by the **solar-powered robotic lawn mower** (USA). The American George Ballas got the idea of a **string trimmer** (strimmer) while watching a car wash in 1971. The powered **rotary hoe** (rotovator/cultivator) was conceived in Australia (1912). The concept of a **mechanical hedge trimmer** goes back to 1854 (USA), but a **handheld** (and hand-powered) machine was not available until 1922, with **electric** and **petrol** models following in 1940 and 1955 respectively (all UK). Other labour-saving devices include the **snow-blower** (1925, Canada) and **leaf-blower** (1950s, USA). The idea of a **snow-maker** dates from 1950 (USA). Those less inclined to get active might always choose to sit back in a **deck chair** (1850s, UK/USA) or perch on a **shooting stick** (nineteenth century, UK) and browse through the first **book about gardening** written in English: Thomas Hill's *The Profitable Arte of Gardening* (1563, UK).

COMMUNICATION

DRUMS, BIRDS AND BEACONS

The earliest form of long-distance communication was probably beating a **drum**; the oldest of these instruments, found in Antarctic ice, is reckoned to have been made around 30,000 years ago. **Signal flags** developed from emblems carried by early armies *c.* 2500 BC (Iraq), and it's thought signal **beacons** were also first employed around this time. **Carrier pigeons** certainly carried the results of the original Olympic Games (776 BC onwards, Greece), and **smoke signals** appear to have been used in China around the same time (eighth and ninth centuries BC). A **mechanical telegraph** system was constructed in Ireland in 1767, and the first national **semaphore system** – signalling by pivoted arms on a post – was set up in France in 1792.

ELECTRICAL SIGNALS

Electrical signalling began when the Danish physicist Hans Christian Ørsted showed how an **electric current** could move a **magnetic needle** (1820). This led to the **telegraph** of Joseph Henry (USA) and Sir Charles Wheatstone in 1831 (UK), and the simpler version of Samuel Morse (1836, USA) who also devised the **code** that bears his name. Next came speech down the wire with Alexander Graham Bell's invention of a practical **telephone** (1876, USA and UK). It paved the way for the **phone book** (1878, USA), the **public call box** (1881, Germany – the same year as the first **international call**, USA to Canada), the **pay phone** (1889, USA), the **automatic telephone exchange** and **dial phone** (1891, both USA), the **answering machine** (1898, Denmark,

The first commercial, portable mobile phone, the DynaTAC 8000X, 1984

though a commercially successful version was not available until 1949, USA), **directory enquiries** (1906, USA), the **video phone** (known as the *ikonophone*, 1927, USA), and in 1933 the **speaking clock** (France). The **fax** machine and **telegram** go back to the work of the Scotsman Alexander Bain who patented an Electric Printing Telegraph in 1843;

Xerox introduced the modern **commercial fax machine** in 1964 (USA). In 1973, Motorola (USA) demonstrated the first mobile phone handset; weighing a staggering 1.1 kg, with 30 minutes talk time, it took 10 hours to charge. The same company produced the first model sold to the public: the DynaTAC 8000X in 1983.

RADIO

In 1865, James Clerk Maxwell (UK) predicted the **existence of radio waves**, and by 1888 Heinrich Hertz (Germany) had **proved their existence**. They were first **transmitted** in 1894 (Oxford, UK), and the following year Guglielmo Marconi (Italy) produced the first **radio transmission system**. Roberto Landell de Moura, a Brazilian priest, made the first **voice transmission** in 1900. A message was sent **across the Atlantic** in 1901 (USA/UK), and the first **radio broadcast** followed on Christmas Eve, 1906 (USA). It featured the first **broadcast music**, a recording of a piece by Handel with the broadcaster himself, Reginald Fessenden, playing the violin. Later came a **news broadcast** (1920, USA), a **weather forecast** (1921, USA) and the BBC's iconic **shipping forecast** (1924, UK). A portable valve radio was made in 1924 (USA) and **push-button tuning** was launched in 1926 (USA). The first successful **transistor radio**, the Regency TR-1, went on sale in 1954 (USA), and Trevor Baylis invented the wind-up, **clockwork radio** in 1991 (UK). **Transmission technology** went from **AM** in 1906 (USA), to **FM** (patented in 1933, USA) and **digital** (1995, Norway). The **BBC**, perhaps the world's best-known radio broadcaster, took to the air in 1922.

TELEVISION

The word '**television**' (**TV**) was coined by the Russian scientist Constantin Perskyi in 1900, and in 1925 John Logie Baird gave the first demonstration of a **mechanical TV** in action (UK). A **transatlantic TV signal** was sent in 1928 (UK/USA), the same year the world's first **TV station** – WGY Television – opened in the USA. Baird made the first **outdoor TV broadcast** in 1931 (UK). The next step, **electronic TV**, became possible thanks to Karl Braun's invention of the **cathode ray tube** (1897, Germany). The **electronic TV receiver** was of Japanese manufacture (1926), and an **all-electronic TV system** was made public in the USA in 1934. Two years later, the first **regular electronic TV broadcast** began in London, UK. The Russians inaugurated **625-line**

A Bush television, c. 1952

TV in 1944 and a **LED TV screen** was demonstrated in the USA in 1978. Baird had demonstrated **colour TV** as early as 1938 (UK), but World War II hampered further progress and the first **national broadcast in colour** had to wait until 1954 (USA). The US also pioneered **digital TV broadcasting** in 1994 but not the **smart TV**, which was developed by Samsung (South Korea) in 2008. **Cable TV** was launched in the US states of Arkansas, Oregon and Pennsylvania in 1948, and NASA launched the first **TV satellite** (Telstar) in 1962. In terms of **programming**, these firsts established enduring genres: regular news broadcast: Lowell Thomas (NBC, USA) in 1940; soap opera: and children's TV: *Children's Hour*, (BBC, UK) in 1946; *These Are My Children* (NBC, USA) in 1949.

RECORDING

Sound recording began in 1877 with Thomas Edison's **phonograph** playing back 'Mary had a little lamb' from a strip of tinfoil (USA). Recordings on **discs** were first made in 1887 (Germany and USA) and the popular material for these was **shellac** (1895, USA) spinning at **78 rpm** (standardized worldwide in 1925). **Electrical** rather than acoustic **recording** started in 1925 (USA) and RCA Victor launched **vinyl-based discs** in 1931, ready for the **12-inch LP** (Long Play) disc in 1948: Columbia's Microgroove. The same company issued the 7-inch **45** the following year. The **record changer**, allowing discs to be stacked for consecutive playing, was devised in Australia in 1925. **CDs** (compact discs) were released commercially in 1982 (universal). An alternative way of recording sound, using **magnetic tape**, had been invented in Germany in 1928. Initially, large **reel-to-reel** machines were used. These were supplemented by the **endless loop**

cartridge (1954, USA), the **cartridge cassette** (1958, USA) and the Philips **compact cassette** (1962, Netherlands). **Stereo** had been available since its invention by the British engineer Alan Blumlein of EMI (1930–1). **Videotape** was first seen in 1951, and Sony introduced a **commercial version** in 1969 (Japan). In the 1990s, these methods of recording and playback were largely superseded by **digital technology**.

THE INTERNET

The possibility of internet communication started with the concept of **packet switching**, conceived in 1961 (USA) and named in 1965 (UK). In the same year, two MIT (Massachusetts Institute of Technology) **computers communicated** with each other by packet switching (USA). **Email** was launched in 1972 (USA), and the UK and Norway inaugurated **international communication** in 1973, the same year as the term '**internet**' was adopted. A **common protocol** for communication was agreed internationally in 1974–8, and '**cyberspace**' coined in the USA in 1984. On these foundations, the first **ISP** (Internet Service Provider) was created in 1974 (USA), **online banking** launched in 1981 (USA), **domain names** appended in 1983 (international), the first **commercial router** sold in 1987 (USA), and **HTML** devised to inaugurate the **World Wide Web** in 1989–91 (UK). In 1992, online **audio** and **video** became available, **smartphones** went on sale, and the phrase '**surfing the web**' was invented (all USA). **Webcams** followed in 1993 (USA).

Other significant debuts, all from the USA, include the **Windows operating system** (1985), the **Microsoft web browser** and use of the term '**social media**'(both 1994). These were followed by shopping on **Amazon** and **eBay** (1995), the inaugural **Netflix** programme (1997), **Google** (1998) and **Wikipedia** searches (2001), and the first **Skype** link-up and use of the **Android operating system** (both 2003). **Facebook** pages popped up in the next year, then the first **YouTube** videos (2005), **Twitter** messages (2006), **iPhone** calls (2007), and **Instagram** images (2010). Amazon's **Alexa** started answering your questions in 2014.

PHOTOGRAPHY AND FILM

TAKING PICTURES

The first photographic process was **heliography**, invented by the Frenchman Nicéphore Niépce (see p.149) in 1824. His work was carried on by a colleague, Louis Daguerre, who devised the **daguerreotype** in 1838, and then by William Fox Talbot, who introduced the **negative-positive** process (1841, UK). Before this, Sarah Anne Bright had become the first **woman** to create photographic images (1839, UK). Various chemical experiments and discoveries led to **celluloid film** (1887, USA). Thomas Sutton (UK) made the **single-lens reflex camera** (SLR) in 1861, and the earliest **easy-to-use camera**, the Kodak box, went on sale in the USA in 1888. The famous **Kodak Brownie** camera followed in 1900, price $1.00. **Polaroid instant cameras** were marketed in 1948 (USA), the **fully automatic camera** (Agfa, Germany) in 1959, and a camera with **through-the-lens metering** (Nikon, Japan) in 1962. The same company produced an **underwater 35 mm camera** the following year. Film had advanced, too, from the first **colour pictures** by Edmond Becquerel in 1848 (France), to **120 format** (1901, Kodak, USA), **35 mm format**

in a cartridge (1934, USA) and **colour prints** (1942, Kodak, USA). But Kodak, the company at the forefront of traditional photography for so long, dramatically helped hasten its downfall with the **megapixel sensor** in 1986 and the **photo CD** in 1992 (both USA). Popular **digital cameras** went on sale in 1995 (USA), and three years later the first patent was taken out for a **mobile phone with a camera** (USA). Japan's J-SH04 was apparently the first **mobile phone capable of taking and sharing pictures**.

THE FIRST MOVIE

The title 'first movie' is controversial. Was it the series of images of a galloping horse, taken at the rate of twenty-five per second using a row of cameras operated by tripwire and shown by the Englishman Eadweard Muybridge in 1878 on the revolving drum of a zoetrope (a device invented in 1834, UK)? Or was it Frenchman Louis Le Prince's famous 2.1-second Roundhay Garden Scene of 1888? Or the flickering pictures produced in 1891 by William Dickson's 'kinetoscopic' motion picture camera (UK and USA)? Or the French Lumière brothers' fifty-second *The Arrival of a Train* (1895) that apparently had the audience screaming and rushing to the back of the room to avoid the oncoming engine? Or the same brothers' famous *Workers Leaving the Lumière Factory* (*Sortie de l'usine Lumière de Lyon*), also 1895? Whichever you choose, *The Story of the Kelly Gang* (1906, Australia) was unquestionably the first feature-length multi-reel film.

MAKING AND SHOWING MOVIES

The Edison Corporation (USA) set up the first **movie studio** in 1894, and **Hollywood** shot its first moving picture in 1910. **Cutting and editing** were pioneered in 1903 by the American Edwin S. Porter (USA), who was also responsible for the first **Western**. Berlin (1895, Germany) and Pittsburgh (1905, USA) both claim to have hosted the first **cinema**. **Home movies** became feasible in 1912 (USA). The earliest **colour feature film** – *The World, the Flesh and the Devil* – was made two years later. *Becky Sharp*, the first truly successful colour film, was screened in 1935 (Technicolor, USA). *Fantasmagorie* (1908, France) launched the **animated cartoon**. **Disney** (USA) entered the market in 1924, and made *Snow White and the Seven Dwarfs*, the first **full-length animated feature**, in 1937. The first **full-length talkie**, *The Jazz Singer* (USA), had been shown ten years earlier, though Thomas Edison's **kinetophone** (USA) had made

A Walt Disney Productions poster for *Snow White and the Seven Dwarfs*, 1937

talkies feasible since 1910. A **3D movie** was shown in 1922, and widescreen **CinemaScope** introduced in 1953. The introduction of **2D CGI** (Computer Generated Imagery) came in 1973, followed by **3D CGI** in 1976, and *Toy Story*, the first **feature-length animation made entirely by CGI**, in 1996 (all USA). Moral guidelines for movies shown to the American public began with the **Motion Picture Production Code** of 1922, which was replaced by the **film rating system** in 1968. The **British Board of Film Censors** (1912) was renamed the **British Board of Film Classification** in 1984. The Canadian Florence Lawrence (1886–1938) is often called the first **movie star**. The **Academy Awards** (Oscars, USA) were launched in 1929, and the **Cannes Film Festival** (France), with its prestigious *Palme d'Or* (Golden Palm, 1955), started in 1946.

MEASUREMENTS, DEVICES AND CLOCKS

NUMBERS

The first known attempt to record how many of something there were – i.e. **counting** – may be either the twenty-nine notches on a 44,000-year-old baboon bone found in South Africa or the Czechoslovakian Wolf Bone of *c.* 28,000 BC which bears fifty-five scratched marks. **Place value** (i.e. composing large numbers by arranging smaller numbers in order, e.g. 1 followed by 2 is 12, etc.) began with the Mesopotamians (*c.* 3400 BC, Iraq), while the Egyptians were the first to work with a base-ten (**decimal**) system *c.* 3100 BC. They were also using **fractions** by 1800 BC and had **zero** (represented by a symbol meaning 'beautiful') as a placeholder by *c.* 1770 BC. Brahmagupta, an Indian mathematician, first gave zero numerical value in AD 628. **Infinity** was documented in India *c.* 1100 BC, and the Chinese were recognizing **negative numbers** (minus numbers) by the first century BC. The Babylonians (Iraq) came up with **square roots** *c.* 1700 BC, leaving the ancient Greeks to give us **percentages**, **prime numbers** and a host of other mathematical concepts and theorems. Our current system of **numerals** (1, 2, 3 etc.) was born in India around AD 200. The Babylonians and Egyptians were the first to give **written recognition of pi** (π), and the Greek genius Archimedes (see p. 124) was the first to calculate it fairly accurately. William Jones was the first person to employ the **π symbol** (1706, UK).

UNITS OF LENGTH

Early recorded **units of length** related to the adult human body – the finger, hand and arm. The Egyptians, ancient Indians and Mesopotamians (Iraq), for instance, were using the **cubit**, roughly the distance from the elbow to the tip of the middle finger, around 3000 BC. The unit preferred by the Romans, Greeks and cities of the Indus Valley (Pakistan) was the **foot** (first millennium BC). One thousand ancient Roman paces gave us the **mile** (first millennium BC), and the **league** – the distance walked in an hour – dates from around the same period. The English clergyman Edmund Gunter (1581–1626) proposed the **nautical mile** as one minute (or one sixtieth of one degree of latitude), while the Chinese believe their equivalent of a mile, the ***li***, was established *c.* 2600 BC. The **fathom** (approximately six feet of depth) goes back to an ancient Greek measure, the *orguia*.

The **metre** was first suggested by another English clergyman, John Wilkins, in 1668, and established in France as the **international decimal unit of measurement** (supposedly one ten-millionth of the distance between the North Pole and the Equator) in 1791.

UNITS OF VOLUME

Units of volume varied widely with time and place, making it impossible to ascertain clear 'firsts'. The following Sumerian measurements of around 2500 BC are as early as any; as their names suggest, they were units of volume as well as weight: **bowl**, **vessel** and **bushel** (the English 'bushel', derived from the old French, was first used in the late eleventh or early twelfth century). **Gallon**, initially meaning a bowl(-ful), and **pint** (one eighth of a gallon, the idea of dividing into eight was apparently of Roman origin) both originated in early medieval Europe. The **hogshead** was defined as about 50 gallons in the fifteenth century (UK), while the **tun**, meaning a barrel of about 250 gallons, dates back at least to the twelfth century (UK and France). Revolutionary France tried to sort out the muddle of global measurements by establishing the **litre** (one cubic decimetre, originally called a 'cadil') in 1795.

UNITS OF WEIGHT

Revolutionary France also attempted to standardize the whimsical world of weights when it established the **gram** (or gramme – the weight of a cubic centimetre of water) in 1795. Before that, weight had been recorded in all sorts of ways, starting with common objects such as stones and seeds. The jeweller's **carat**, for example, was originally the weight of a carob seed. The **pound** comes from the Latin *libra*, which is why it is abbreviated as 'lb', and the **ton** was the weight of a large barrel or 'tun'. The UK standardized its **imperial system of weights and measures** in 1824; the US Treasury took a first step towards doing the same in 1832, but the **US customary units** were officially standardized only in 1893. Worldwide agreement on a **uniform system of metric units of measurement** was achieved in 1948.

MEASURING

The first known **ruler** – an instrument for drawing straight lines and measuring, rather than a potentate in a crown – was a 4,650-year-old copper instrument from Assyria (Iraq/Syria). The German merchant Anton Ullrich invented the **folding ruler** in 1851, and the American Frank Hunt came up with a **flexible ruler** in 1902. **Tape measures** began with ancient Egyptian 'rope-stretchers' who measured out agricultural land beside the Nile after the annual floods had receded (third millennium BC). A **retractable measuring tape** in a case was patented in 1868 (USA). Scales for weighing seem to have originated in the Indus Valley (Pakistan) *c.* 2000 BC. The more accurate **Roberval balance** was made in 1669 (France), and **spring scales** were first made around 1770 (UK). Even more precise were the **quartz crystal microbalance** of 1959 and, for general use, the **digital scales** of *c.* 1980 (both USA).

CALCULATING AND COMPUTING

The first calculator, the **abacus**, was made in Mesopotamia (Iraq) *c.* 2500 BC, while the Greek Antikythera mechanism (*c.* 100 BC), found on the seabed in 1901, has been called

A Z3 computer, 1941

the first **analogue computer**. Following the invention of **logarithms** (1614, UK), William Oughtred (1574–1660, UK) devised the **slide rule** (and the **X symbol for multiplication**). The German Wilhelm Schickard made the first **adding machine** in 1621, though not until the **Arithmometer** in 1820 (France) was there a machine capable of performing all four mathematical functions. Charles Babbage's 'Analytical Engine' conceived in 1834, a century ahead of its time, was the first **programmable computer**. **Push-button calculators** appeared in 1902, and a machine able to **solve equations** in 1921 (both USA).

Then came the electronic revolution, which many say began in 1937 when Alan Turing conceived of a **theoretical computing machine** (UK). This was followed by Konrad Zuse's Z3, the first **programmable computer** (Germany, 1941), the first **integrated circuit** (silicon chip, USA, 1958), **computer**

game (*Spacewar*, USA, 1962), **word processor** (IBM, USA, 1964), **RAM and microprocessor** (Intel, USA, 1970), and in 1975 the first **personal computer** or PC (Altair, USA).

UNITS OF TIME

It's odd, isn't it, that at a time when almost all our other measurements are decimal, we still have days of twenty-four hours and years of twelve months. For the former we need to go back to the ancient Egyptians, the first to divide the **day into twenty-four parts** (third millennium BC). The Babylonians, whose mathematical system was based on the number sixty, divided each **hour into sixty minutes** and each **minute into sixty seconds** (*c.* 3500 BC). (We have also retained their division of a circle into 360 degrees, but not their sixty-hour day.) A **year of twelve months** comes from the moon, perhaps beginning *c.* 4000 BC in the Fertile Crescent (see p.7). The **solar calendar** dates from Julius Caesar (*c.* 45 BC, Italy) and our present **Gregorian calendar** from 1582.

SHADOWS AND SAND

The most obvious way of telling the time of day – by looking at the sun – gave rise to the first timepiece, the **sundial** in Babylon (Iraq) and Egypt (*c.* 1500 BC). At about the same time, these civilizations also invented **water clocks**, though some argue that the Chinese had devised a water clock long before this. The **hourglass**, which works on the same principle, was probably of Roman devising *c.* AD 350. Using a **candle as a timepiece** is attributed to the Chinese (*c.* AD 520), as is the first **hydraulic mechanical clock** (AD 725).

PRECISE TIME

There is much debate over the invention of the **escape mechanism** that gave rise to the **all-mechanical clock** with its distinctive ticking; some argue for the French artist Villard de Honnecourt (*c.* 1237), while others suggest the honour should go to the makers of Dunstable Priory clock (1283, UK) or a clock in the Visconti Palace, Milan (1335, Italy). The next development was the **spring-driven clock** (**clockwork**) made in fifteenth-century Europe, leading to the **pocket watch** (sixteenth century, Europe), and **minute hand** (1577, Europe). A timepiece given to Elizabeth I of England in 1571 is the earliest mention of a **wristwatch**. In 1776 the **stopwatch** was devised in France, and a Swiss watchmaker produced the **self-winding watch**. Between 1735 and 1761, John Harrison (UK) worked painstakingly to produce the first practical **marine chronometer** that allowed sailors to ascertain their longitude with great precision. The first **electric clock** was made in 1840 (UK); the Rolex Oyster was the first genuinely **waterproof wristwatch** (1926, Switzerland); and in 1972 another Swiss company, Hamilton, manufactured an **electronic watch**. In 1927, the Canadians made the first **quartz clock**, though Seiko (Japan) did not produce the first **quartz wristwatch** until 1969. The ultimate in timekeeping, the reliable **atomic clock**, was created in the UK's National Physical Laboratory in 1955. Finally, in order to coordinate all these clocks and watches, the International Meridian Conference of 1884 agreed **Greenwich Mean Time** (UK; superseded in 1972 by Coordinated Universal Time) as the single global time axis.

THEORY

SUBJECTS

In the earliest academies of the East and West, **philosophy** was the first (and most important) subject (*c.* 500 BC). **Mathematics** was there at the start, too, and **music** was regarded as good for the body and soul. The Romans favoured **engineering**. In the Christian-Muslim Middle Ages, **theology** (religion) became a sine qua non, with European universities also supplementing the subjects of the ancients by dividing mathematics into **arithmetic** and **geometry**, and adding **astronomy**, **music theory**, **grammar**, **logic**, **astronomy** and **rhetoric**. The more forward-thinking institutions even included **medicine**. Knowledge of **Latin** and **Greek** was taken for granted. In Europe, **law** was accepted as an academic subject in the Middle Ages; musical institutions, including the **conservatoires**, emerged in sixteenth-century Italy (see p.246). Mathematics was embracing **physics** by the end of the seventeenth century, and **veterinary medicine** was taught in France from 1761. Modern **historical scholarship** is said to have begun with the Tunisian scholar Ibn Khaldun (1332–1406), but it was not until the nineteenth century that disciplines such as **history**, **biology**, **chemistry**, **geography**, **economics**, **accountancy**, **modern languages** and their **literature**, were subjects in their own right. All the other subjects available today, from **aeronautical engineering** to **youth work**, are creations of the twentieth or twenty-first centuries.

PATENTS, COPYRIGHT AND PRIZES

The dating of many of the more recent 'firsts' is based on when the inventor applied for or was granted a **patent.** The origin of the patent idea – granting exclusive rights to a creator – can be traced back to a regulation issued in the notoriously dissolute Greek city-state colony of Sybaris to protect discoverers of new luxuries (500 BC). More recently, King Edward III of England issued monopoly-granting '**letters patent**' to help new industries (1331). The idea caught on, and in 1555 France's Henry II was the first to insist on a **written description of an innovation**. Patents went **international** with the Patent Cooperation Treaty of 1970. **Trademarks** may have been initiated by the sword-makers of the Roman Empire (first century AD, Europe), but they did not receive **legislative backing** until 1266 (UK). France issued the first **comprehensive trademark legislation** in 1857. The first **copyright law** was issued in Britain in 1710, and the idea was **extended worldwide** by the Berne Convention of 1886. **Nobel Prizes**, the most prestigious awards primarily for those working in science and engineering (specifically chemistry, physics and physiology/medicine, but also including literature) were established in 1895 (Sweden). Other major international science awards include the **Albert Einstein Award** (theoretical physics, 1951, USA), the **Albert Einstein World Award** of Science (1984, World Cultural Council), and UNESCO's **Kalinga Prize** for the Popularization of Science (1952).

GUILDS AND UNIONS

STRENGTH IN UNITY

Wherever skilled craftsmen and merchants were grouped together in an urban setting, they formed associations – commonly known as **guilds** – to protect their interests. The first guilds appear to have been Chinese (*c.* 1000 BC), with Roman ***collegia*** the first in Europe (*c.* 200 BC). After the fall of Rome, a different type of European guild emerged. Its members pooled their gold (hence 'guild') for safekeeping, and swore oaths to back each other in all circumstances and seek out common enemies – not unlike today's street gangs! Better-known **craft and merchant guilds** (cloth merchants, silversmiths, etc.) were up and running by the twelfth century. To some extent they morphed into the early **unions**, organizations of skilled workers, like the group calling itself the Journeymen Hatters of Great Britain and Ireland (1667) or Exeter's Company of Weavers, Fullers and Shearmen (early eighteenth century, UK). But the first **union in the modern sense** – an association of employed workers acting together to further their mutual interests – belonged to the industrial age. Though these early unions were small, scattered and weak, leaving little record of their existence, we know they existed because they were outlawed by the British government's **Combination Act** of 1799. The General Union of Trades (aka the Philanthropic Society) was the first attempt to bring together **workers of different trades** (1818, UK); the National Association for the Protection of Labour (1830, UK) was a prototype **national general trade union**.

STRIKES AND LEGISLATION

A possible first **strike action** was when those working on the Royal Necropolis at Deir el-Medina rose up against Pharaoh Ramses III in 1152 BC. Later instances of withdrawal of labour are too numerous and ill-documented to identify a first instance. The refusal of all the *plebs* (ordinary people) to run the city of Rome in 494 BC might be considered the first **general strike**. The first modern instance of a widespread withdrawal of labour, involving almost 500,000 workers, occurred in England in 1842. The Plug Riots, as they were called, followed parliament's rejection of a petition for widespread political reform and were sparked by discontent in the coal mines and cotton mills. It is impossible to say when strikes were first prohibited because throughout history virtually all governments, local and national, have taken steps to limit or repress such actions and the organizations behind them.

THE ENVIRONMENT

EARLY WARNING

The first **significant warning** about the fragility of humanity's place on Earth was made by Thomas Malthus in *An Essay on the Principle of Population* (1798, UK) in which he predicted that **population growth** would inevitably exceed our ability to feed everyone. At the same time, the Romantic movement was urging a greater **respect for the natural world**. In 1852, the Englishman Robert Smith talked of '**acid rain**' and analysed its causes and effects. '**Ecology**' was coined in 1866 (Germany), and in 1896 a Swedish chemist introduced the idea of the **greenhouse effect** of pollutants in the atmosphere. The first death from **asbestosis** was confirmed in 1924 (UK). We started reading about '**climate change**' in 1952 and '**global warming**' in 1957 (both USA). Rachel Carson's *Silent Spring* (1962, USA) was the first popular **book** to bring home the seriousness of environmental issues. Five years after its publication, the first **mega-tanker disaster** struck when the *Torrey Canyon* was wrecked off the coast of Cornwall, UK. A massive hole in the **ozone layer** was identified in 1985; two years later came a warning linking global warming to the possibility of the Gulf Stream failing and Europe entering a **new Ice Age**. In 1992, the Intergovernmental Panel on Climate Change (IPCC) issued its **first report**, stating that human activity might well **raise global temperatures** by 0.3°C per decade. The sinking of Greenpeace's *Rainbow Warrior* by the French intelligence services was the first known **significant use of force against an environmental protection organization** (1985).

ACTIONS, ORGANIZATIONS AND AGREEMENTS

Human attempts to mitigate or even reverse the escalating environmental crisis have taken several forms.

Pollution

As the first industrial nation, and therefore the first to suffer the ill effects of urban and industrial pollution, the UK passed the first **significant nationwide environmental protection laws** with the Public Health Act of 1848 and the Alkali Acts of 1863 and 1874. **Global oceanic pollution** was accepted as a problem when the Inter-Governmental Maritime Consultative Organization (later the International Maritime Organization) was established in 1948. **DDT** was widely banned in 1970, and **chlorofluorocarbons (CFCs)** in 1987. In the twenty-first century, the pace of environmental protection action accelerated as the warnings of scientists grew ever more dire. In 2002, Bangladesh led a global movement to **ban the use of plastic bags**. In 2009, the year Ireland became the first nation to **tax plastic bags**, Japan launched the first **satellite to monitor carbon dioxide emissions**, and the Australian town of Bundanoon banned the sale of bottled water. Norway undertook to **phase out all conventionally powered cars** by 2025 (2016), and in 2017, Paris, Madrid, Athens and Mexico City pledged to be the first **large cities to outlaw diesel vehicles**. (See also organizations, p.190)

Conservation and wildlife

India led the way in serious **environmental conservation**, first with the Madras Board of Revenue's local forest protection scheme (1842), then with the world's first **permanent and large-scale forest conservation** programme (1855). The first **national fauna-protection law**, the Sea Birds Preservation Act, was passed in 1869 (UK). The Plumage League, the pioneer **organization for the protection of wildlife**, was set up in 1889, quickly developing into the Society for the Protection of Birds (1891, UK). Yellowstone (1872, USA) was the world's first **national park**. England's National Trust (1895) developed into the first **NGO to coordinate all conservation efforts**. Whales received their first protection with the formation of the **International Whaling Commission** (1948), the same year as the establishment of the **International Union for the Protection of Nature** (renamed the International Union for Conservation of Nature, 1956). The **World Wildlife Fund** (WWF) was founded in 1961, and in 2014 the UN Environment Programme (UNEP) launched its first **World Wildlife Day**. (See also organizations, p.190.)

Resources

The French professor Augustin Mouchot issued the first **scholarly warning of the finite nature of Earth's resources** in 1873. Safe water was the first to come under strain, resulting in the first **World Water Day** in 1993. Eighteen years later, the UN's first **assessment of Earth's resources** reckoned that one quarter of all agricultural land was 'highly degraded'. The year 1983 saw the first **genetically modified** (GM) **plant** (USA), and in 1994 the first GM food – the Flavr Savr tomato – went into production (USA). Golden rice was the first plant **genetically modified to improve its nutrient value** (2000, USA). (For the switch to green energy, see p.190)

Legislation and agreements

Denmark became the first country to have a **cabinet minister** whose exclusive brief was **to handle environmental issues** (1971). The next year, the UN held its first **Conference on the Human Environment** (Stockholm, Sweden), leading to the **UN Environment Programme**. The 1992 conference led to the **Framework Convention on Climate Change**. **Green politics** began with success in local elections for candidates with environmental issues on their platform (1970, Netherlands). The United Tasmania Group (Australia, 1972) was the **inaugural Green political party**; New Zealand's Values Party was the first to **contest parliamentary**

seats nationwide, and the German Greens (twenty-seven Bundestag seats, 1983) were the first to have a **significant influence on global politics**. Under the terms of the 1997 Kyoto Protocol (Japan), the nations of the world agreed to **implement broad outlines of control on the emission of harmful greenhouse gasses** by 2005. The 2015 Paris Agreement produced international consensus for the first time to limit global temperature to no more than 2°C above pre-industrial levels. In the same year, Switzerland was the first country to make a clear **pledge to halve its greenhouse gas emission** by 2030.

Organizations

The Sierra Club (1892, USA) and the Coal Smoke Abatement Society (1898, UK) vie with each other for the crown of the

An Extinction Rebellion climate change protest, 2019

first environmental NGO (Non-governmental organization). Other foundations (excluding those already mentioned) and movements include Friends of the Earth (1969), Greenpeace (1971), and Extinction Rebellion (2018).

PART VI: PEACE AND WAR

GOVERNMENT

CITIES AND RULERS

The first political organization, dating back hundreds of thousands of years, must have been the **tribe** or, when the members were somehow related, the smaller **clan**. **Scottish clans**, some of the best known, are steeped in myth, though those able to trace their line back to the fifth-century Irish King Niall of the Nine Hostages (e.g. Clan MacNeill) may have a valid claim to be the first. The earliest **political institutions**, stemming from the development of agriculture (see p.7), were most probably the **city states** of the Sumerian civilization (*c.* 4000 BC, Iraq), notably Uruk. Sumer also provides the first verifiable **monarch**, King Enmebaragesi of Kish (*c.* 2600 BC, Iraq), and the earliest evidence for **slavery**. The first known **woman ruler** may have been the Egyptian queen (or pharaoh) Merneith, who took over the reins of power on the death of her husband around 2950 BC. Sobekneferu of Egypt (1806–1802 BC) was the first **historically verifiable queen**, though there is a possibility that Enmebaragesi was female, and an argument can be made for Queen Ku-Bau of Sumer (*c.* 2400 BC). Egypt hosted the first recorded **coronation**, that of Pharaoh Seti I (*c.* 1290 BC).

FORMS OF STATE

The world's first **state**, as opposed to a state comprising a single city or league of cities, was probably Egypt, which was united around 3150 BC by a ruler – mythical or otherwise – known as Menes or Narmer. It is not possible to ascertain the earliest **country**, because the word is loosely used and generally refers to a geographical area. As the term is open to different meanings, the **nation state** is equally controversial; claims for the first are made by Armenia (eighth century BC), Japan (seventh century BC), Iran (sixth century BC), and in more recent times, Scotland (1320) and fifteenth-century France and England. Great Britain boasts the first official **national anthem** – 'God Save the King/Queen' (1745) – and Denmark's white cross on a red background was the first official **national flag** (*c.* 1307). While the US was the first constitutionally **federal state** (1787), the idea of a **republic** governed by popular sovereignty goes back to Greece in the eighth century BC or the Indian city state of Vaishali (seventh century BC).

RULERS

Britain pioneered the concept of **constitutional monarchy** with its Glorious and Bloodless Revolution of 1688–9. **Limited monarchy** existed long before that, for example where the monarch was elected, as in Rome from 753 BC and Silla (Korea) in 57 BC. The first **throne** – monarch's seat – is as old as the institution itself. Crowns were preceded by royal

A Korean metal crown, originally encrusted with semi-precious stones, sixth century AD

diadems, the earliest known being worn by the priest-kings of the Indus Valley civilization (*c.* 3000 BC, Pakistan); the first **crowns** as we understand them were the tall *hedjet* and *deshret* hats sported by the first rulers of Upper and Lower Egypt (fourth millennium BC). Very early **circular metal crowns** were worn by the early kings of Korea, from around 18 BC. The territories ruled by Eannatum of Sumer (2454–2425 BC) are often called the first **empire**, though Jimmu of Japan is supposedly the first ruler accorded the title **emperor** (*c.* seventh century BC). However, as 'emperor' has linguistic and cultural ambiguities, and Jimmu apparently lived to the age of 126 and was descended from a sun goddess and storm god, the first verifiable emperor may be either Darius of Persia (522–486 BC) or Qin Shi Huang of China (247–210 BC). Less controversially, Suiko of Japan (AD 593–628) was the first **empress.** **Dictators** were initially magistrates and military commanders in the ancient Roman Republic (509 BC onwards); the word acquired a negative connotation from 82 BC. The US was the first country to be headed by a **president** (1787).

SYSTEMS OF GOVERNMENT

As the early kings and queens claimed divine status, the first monarchs (see p.193) were to some extent the first **theocracies.** Egypt's pharaohs (*c.* 3150 BC onwards), for instance, were absolute rulers who mediated between the deities and human beings. China's Shang dynasty (*c.* 1556–1046 BC) was theocratic, but perhaps the first **true theocracy** was the Islamic caliphate established after the death of the Prophet Muhammad in AD 632. (Shia Muslims, however, reckon the true caliphate did not begin until the time of Imam Ali in AD 656). The first **oligarchies** were likely to have been among the city states of Mesopotamia (see p.193). **Aristocracy**, originally meaning government by the best, came to mean rule by a privileged class – as in Athens and other Greek city states in the seventh century BC. The same states were among the earliest **plutocracies**, too, as the privileged used their position to accumulate wealth as well as power. Ancient Athens is often hailed as the first **democracy**, although its women and slaves would not have agreed. By giving all citizens the right to vote and run for office in 1906, Finland probably became the first **true democracy**. However, as it was technically a Grand Principality of Russia, maybe the prize should go to Denmark (a constitutional monarchy) in 1915.

CONSTITUTIONS, MINISTERS AND OFFICIALS

While law had been written down before his time, Aristotle (see p.213) was the first person to talk of a **written constitution**. The Roman Republic then borrowed Greek ideas to formulate the Law of the Twelve Tables (450 BC), a codification of laws with constitutional elements. Japan's Seventeen Article Constitution of AD 604 was the first such document based around moral principles, leaving the English republic's Instrument of Government (1653) as the first **full and detailed written constitution**. **Ministers** sat in the first **councils** of earliest Mesopotamian monarchs (see p.193), and there is no reference to a **cabinet** before 1605 (UK). While many monarchs had chief ministers, the title of **prime minister** is generally said to date from the time of Britain's Robert Walpole (in power 1721–42), with Sirimavo Bandaranaike of Sri Lanka (elected 1960) the first **female prime minister** and **elected female head of state**. The earliest reference to a **governor** comes from the Mesopotamian city of Larsa around 2000 BC; **magistrate** dates from the time of the ancient Kingdom of Rome, where the king himself was the chief *magistratus* (753 BC onwards), and **sheriffs** were shire (county) reeves (royal officials) in early eleventh-century Anglo-Saxon England.

PARLIAMENTS AND PARTIES

People have always gathered to discuss problems and policies, and **assemblies** of one sort or another have been a feature of government since the beginning of civilization (see p.7). But the idea of a true **parliament** (a representative gathering rather than an oligarchic council) began with King Alfonso IX of León summoning a Cortes in 1188. **Parliamentary government** grew out of the actions of assemblies in Holland (1581), Britain (1689) and Sweden (1721). The concept of **loyal opposition** (a phrase coined in 1826, UK) – central to any democracy – emerged in Britain during the first decade of the eighteenth century. **Political faction** can be dated to the beginnings of democracy in ancient Athens (see p.196), but faction does not become **party**, a group united by principle and policy, until the England of Charles II (1660–85, see opposite page). The terms '**left**' and '**right**' – radicals/innovators and conservatives – stemmed from the seating arrangements of the French National Assembly meeting in 1789. William Gladstone's campaign of 1878–80 is heralded as the first **modern political campaign**; **election radio ads** were initiated in the US, 1922–3; Dwight Eisenhower (USA) pioneered **TV electioneering** in his 1952 campaign; and Barack Obama did the same with **social media** in 2008.

BUREAUCRACY AND BUDGETS

The invention of writing (see p.237) led to the formation of the first **bureaucracy** in ancient Sumer (see p.197), though the word itself was coined by Frenchman Jacques de Gournay (1712–59). Entry into the civil service by **competitive exam** was established in China by AD 605, while the idea of a permanent, **non-political central administration** was pioneered in Britain in 1854–70. The **Exchequer** (named after the chess-board-like cloth used for computation) emerged in ninth-century England; **national budgets** may date from the actions of Robert Walpole (see p.197) in 1721.

WHIGS AND TORIES

The names of the world's first two political parties were terms of abuse. Those who opposed the absolutist inclinations of King Charles II (r. 1660–85, UK) and the possible succession of his Roman Catholic brother James, Duke of York (after whom 'New York'), were slandered as 'Whigamores' – riotous Scottish Presbyterians. Those supporting the king and the right of his brother to succeed were damned as 'Tories' – Irish robbers. The two names stuck, and the Tory Party (officially the 'Conservative and Unionist Party') endures to this day.

POLITICAL CHANGE

The earliest monarchies (see p.193) rapidly became **hereditary**. In ancient Egypt and Sumer, it seems as if **primogeniture** through the male line was the norm; King Gilamesh of Uruk (*c.* 2800 BC), for example, was succeeded by his son Ur-Nungal, who in turn was succeeded by his son Udul-kalama. The replacement of King Lugal-kitun of Uruk by King Mesannepada of Ur may be the first known

An Iroquois woman, 1927

coup d'état (twenty-sixth century BC, Iraq). An alternative is the presumed first **civil war** that followed the death of Pharaoh Qa'a in Egypt (twenty-ninth century BC) and led to the succession of Pharaoh Hotepsekhemwy of a new ruling dynasty. Though there had been rebellions, revolts and uprisings aplenty throughout ancient history, the first we know as a **political revolution** was the overthrow of the Umayyad Caliphate by the Abbasids in AD 750. England's Glorious and Bloodless Revolution (see p.194) inaugurated the era of **modern political revolution**. The first formal **elections** were held in Athens in the sixth century BC, where the first **secret ballots** were also held. We are told that in the seventeenth century, Canadian **women** of the Iroquois tribal councils had the **same voting rights as men**, and certain Swedish women had the vote in the early eighteenth century. The Kingdom of Hawaii was the first **state to introduce universal suffrage** irrespective of gender (1840, rescinded 1852), with New Zealand (the first of the currently extant states) following suit in 1893.

RIGHTS AND THEORY

Chronologically speaking, the first person we know to have given serious attention to **political thought** was the Chinese philosopher Confucius (551–479 BC). Plato's *Republic* (*c.* 380 BC, Greece) is the first **book devoted to political philosophy**, though under the guise of a search for justice. Certain key political concepts also emerged in antiquity: the **state of nature** (i.e. the world before states and governments) from Mozi (*c.* 470 to *c.* 371 BC, China); the **social contract** from Glaucon (brother of Plato, fifth to fourth century BC), and perhaps even **human rights** (highly dubious, despite being propagated by the United Nations) by Cyrus the Great (*c.* 539 BC, Iran). From ancient India, notably in the pronouncements of Ashoka the Great (*c.* 268–232 BC), came early endorsements of **freedom**, **equality** and **toleration**. Cyrus (see p.243) said he had **abolished slavery**, as did China's Qin dynasty (221–206 BC). A British legal judgement in 1772 declared **slavery illegal**, giving heart to the growing anti-slavery movement worldwide. Vermont was the first region in America to abolish slavery (1777); the first **global**

human rights organization, Anti-Slavery International, was established in Britain in 1839. The earliest modern documents associated with the **concept of human rights** are England's Magna Carta (1215, UK), the Twelve Articles of Germany's rebellious peasants (1525), and the Bill of Rights (1689, UK) which led to the first **generally accepted international code of human rights**, the United Nation's Universal Declaration of Human Rights (1948).

COMPETING PHILOSOPHIES

The English pamphleteer Richard Overton (*c.* 1600–64) and philosopher John Locke (1632–1704) were the founders of **liberalism**. **Capitalism**, as old as business itself, was named in 1854 when it appeared in the novel *The Newcomes* by the British novelist William Makepeace Thackeray. Its progenitors appeared much earlier: 'capital' in twelfth-century Europe and 'capitalist' in the Netherlands in 1633. The idea of **communism** can be traced back to Plato's *Republic* (see p.201), though the word was coined only in 1777 (Victor d'Hupay, France) and popularized in *The Communist Manifesto* (Karl Marx and Friedrich Engels, 1848, Germany and UK). The first **communist state** (though it called itself 'socialist') was established in Russia in 1917. Similarly, **socialist** ideas began with Plato (see p.201), the early Christians (first century AD) and the Persian priest Mazdak (sixth century AD), before the word was coined in 1832 (France). The first **socialist local government** was elected in the Paris Commune (1871), the first **socialist state government** in Queensland, Australia (1899), and the first **socialist national government** in Germany (1919).

INTERNATIONAL RELATIONS

TREATIES AND AGREEMENTS

The settlement of a border dispute between the Mesopotamian states of Lagash and Umma is said to be the first known **treaty** (*c.* 2550 BC, Iraq), while the agreement made around 1259 BC between the Egyptians and the Hittites was the first **peace treaty**. **Marriage treaties** feature prominently in the Amarna Letters of ancient Egypt (1350s to 1330s BC). For thousands of years trade was in the hands of individuals or groups of merchants, so perhaps the journeys of the Chinese official Zhang Qian (d. 113 BC) along what was to become the Silk Road may be regarded as the first **trade mission**, and the subsequent agreements the first **trade treaties** (see the Amarna Letters, above). **Tariff-free trade** had operated widely in the ancient world, but the first formal **free-trade treaty** was the Anglo-French agreement of 1860. **Tribute** and **reparations** merged with each other in the ancient world (see the Amarna Letters, above), and the Achaemenid (Persian) Empire (*c.* 550–330 BC) was one of the first states to organize their regular and efficient collection. The Romans demanded a clear tribute from the defeated Carthaginians in the Treaty of Lutatius (241 BC). **Disarmament** was first discussed internationally at the Hague Peace Conferences of 1899 and 1907; the Treaty of Versailles (1919) paid lip service to **arms reduction**, and the Washington Naval Treaty of 1922 limited the construction of capital warships. **CND** (the Campaign for Nuclear Disarmament, UK) was founded in 1958, and the first **Nuclear Non-Proliferation Treaty** was signed ten years later (USA, Russia, and many other states).

An initial **agreement to reduce nuclear weapons** (START I) was signed in 1991 (USA/Russia). **Truces**, **ceasefires** and **armistices** have existed as long as warfare itself (see p.222), but one of the first formal armistices in modern times was the Armistice of Copenhagen that ended the Count's Feud (1537, Denmark). The waving of a **white flag** to indicate surrender seems to have developed simultaneously in China and ancient Rome around AD 100.

DIPLOMACY

The Amarna Letters (see p.203) are among the first records of formal **diplomatic activity** involving **ambassadors** and **emissaries**. The modern system of permanent **embassy buildings** began in Renaissance Italy (1300 onwards). **Diplomatic immunity** gets an early mention in the epic *Ramayana* (*c.* 2500 BC), and the concept was put on a more formal footing by the teaching of the Prophet Muhammad. Britain gave the earliest legal guarantee of diplomatic immunity in 1709, a principle extended worldwide by the 1961 Vienna Convention on Diplomatic Relations. The idea of the **passport** appears in The Bible (*c.* 450 BC). A little later (first century BC), the Chinese authorities came closer to a true passport by insisting on age, height etc., but a 1414 English Act of Parliament is said to have instituted the first **modern passport**. The holder's photo was first appended in 1876 but was not generally compulsory until World War I (1914–18), while **machine-readable passports** were introduced in the 1980s. France issued the first **ID cards** for workers in 1802, and the Ottoman Empire under Sultan Mahmud II was the first state to make them compulsory (1844).

MONEY

COINS

The first **money** – using cattle and grain – was hardly shopping-friendly (see domestication of animals, p.20), as it involved barter-like exchanges such as, 'I'll pay you five cows for your horse/son/daughter.' Commodity money (precious stuff, such as cattle, shells and salt) was given an equivalent value in precious metals like copper and silver around 2000 BC, so that the first recorded **unit of currency**, the shekel, was a weight of barley and of precious metals. To avoid repeated weighing, the metal's weight was stamped on it – giving us the earliest **coinage**. The Chinese paved the way with triangular pieces of bronze (*c.* 1000 BC); more practical round coins followed around 300 years later (India, China, Mediterranean), before the archetypal **solid, disc-shaped coins** – the first were stamped with the head of a roaring lion – were made in Lydia (Turkey) around 2,600 years ago. The Persian Achaemenid Empire, which had conquered Lydia, copied their coinage idea and issued the first **gold coin** about 540 BC. Though it is not easy to distinguish people from deities on early coins, the first **human portrait on a coin** appeared in Lycia (Turkey) around 450 BC, the earliest possibly being the head of the unpronounceable King Teththiweibi. The Swiss produced the first Western coin **dated with Arabic numerals** in 1424. The silver **penny**, derived from the shekel, emerged around AD 600 in England, where 240 of them weighed one **pound**. In 1180, pennies were called '**sterlings**' – and the name remains in today's pound (of) sterling(s). The first **dollar** originated as a 'thaler' (1519, Germany); Japan implemented the silver **yen** (meaning 'round object') in 1871; and the Chinese **yuan** has been around for some 2,000 years. Russia pioneered the earliest **decimal currency** when it divided its rouble into 100 kopecks in 1704. The **euro** entered the market in 1999.

NOTES, CHEQUES AND INSURANCE

The idea of **paper money** began with notes (often on clay tablets), confirming that goods had been left in a storehouse (in various ancient civilizations, *c.* 3000 BC onwards); these notes acquired value in themselves. **Banknotes** originated in seventh-century China, where **true paper money** was in use by the eleventh century. Primitive **banking** was in operation with early trade (see p.203), and some scholars suggest the first **bankers** were members of the Babylonian House of Egibi (*c.* 1000 BC, Iraq). Specific **written reference to banking** comes first from ancient Greece *c.* 400 BC. The earliest known **foreign exchange contract** (1156) was made in Genoa, the year before the first **state-backed bank** opened in Venice (both Italy). The Bank of England (1694) was the first to launch **permanent banknotes**, promising to pay the bearer a certain sum. They were soon being **printed** (rather than handwritten) and the Royal Bank of Scotland pioneered the **overdraft** in 1728. Paying money to someone bearing a signed note (sixteenth century, Netherlands) developed into the **cheque** in seventeenth-century England, where the Bank of England produced the first **printed cheques.** In 1969, several countries issued **cheque guarantee cards**. The first **traveller's cheques** were issued in London in 1772, and three years later a British publican founded the first **building society** in his inn. The earliest evidence of **insurance** is the Code of Hammurabi (see p.81); the first **insurance contracts** were written in fourteenth-century Genoa; and the world's first **insurance company** open to the public was Hamburger Feuerkasse (1676, Germany). There were funeral expenses clubs in ancient Rome, but the first **life insurance policy** was issued in London in 1706.

The plastic revolution began in 1950 with the first Diners Club payment cards (USA). Other initiatives followed: the credit card (1951, USA), the 'hole in the wall' **cash machine** (1967, UK), **debit cards** (1987, USA and UK), **chip and pin** (1992, France), **online banking** (1994, USA), and **contactless payment** (1997, USA).

TAX AND WELFARE

Taxation began in ancient Egypt (*c.* 3000 BC) with the state (i.e. the pharaoh) demanding labour, cooking oil and a proportion of the harvest. Darius I (*c.* 550–486 BC, Persia) established the first **efficient taxation system** based on uniform payments of precious metal, which could be in the form of coins. While pharaonic taxation might be regarded as a tax on income, as might the levies on the better-off demanded by Emperor Wang (AD 10, China), the first universal and progressive **income tax** was levied in Britain in 1798. Based on taxation imposed during World War II and on an experiment in its Ivory Coast colony (1954), in 1958 France introduced **value added tax**. Precedents for **withholding** US **taxes** (**PAYE** – pay as you earn) go back as far as the American Civil War when the US Treasury withheld taxes owed by federal employees under its control (1862). The UK introduced universal PAYE in 1944. Nationwide **rationing** was introduced in Germany in World War I (1914). The concept of the **welfare state** may be traced back to the policies of Ashoka the Great (India, see p.201) and the Rashidun Islamic Caliphate of the seventh century AD, the latter collecting funds for the poor, elderly, disabled and needy. The **modern welfare state** originated in the Germany of Chancellor Bismarck (in office 1871–90), notably with state-funded **old-age pensions** (1889).

JUSTICE

LAW AND LAWYERS

With the emergence of civilization (see p.7), society's rules and punishments for their infringement were written as **law**. This practice may have begun in Egypt 5,000 years ago, though the earliest written **legal code** was that of the Sumerian ruler Ur-Nammu (d. 2094 BC). The first **courts of law** were the royal courts, and monarchs were the first **judges**. Inevitably, royal powers were handed down to viziers, magistrates (see p.196) and other officials who officiated in their own courts. Ascertaining firsts is impossible because they went by different names and had multiple functions – e.g. the Biblical war leader-judge Othniel (late second millennium BC). Amateur orators appealing before the courts of ancient Athens, (*c.* 500 BC, Greece) were the first **lawyers/advocates** we hear of, but the occupation did not become a **profession** until legalized by the Roman Emperor Claudius (AD 41–54, Italy). The Roman Consul Tiberius Coruncanius (d. 241 BC) was the first known **law professor**, and Beirut (Lebanon) possibly had the first **law school** (early third century AD). There was something of the idea of a **jury** in trials by a council of elders for lesser crimes in New Kingdom Egypt (1550–1504 BC) and Vedic India (*c.* 1500 to *c.* 1100 BC), but the modern jury system began with the Germanic tribal custom of investigation and judgement by a group of the accused's peers (first millennium BC), and in the hundred-strong juries of ancient Athens (see above). The modern **jury trial** originated in England around 1166. Though unqualified, Victoire de Villirouët was probably the first **female to act as an attorney** in court (1794, France), and Marija Milutinović may have been the first **qualified female lawyer** (1847, Serbia).

POLICE AND PUNISHMENT

The earliest record of an **organized police force** comes from China in the seventh century BC. Its local 'prefects' included women, arguably the first **female police officers**. A **centralized police force** was set up in Paris, France, in 1667, and extended to other towns and cities thirty-two years later. Napoleon I reorganized his capital's force in 1800, making it the first **uniformed police force**. **Modern policing** – i.e. by a professional force serving the courts and the public, not the state – dates from the establishment of London's politically neutral, unarmed Metropolitan Police in 1829. Its headquarters at 4 Whitehall Place, the site of a former royal palace, backed onto Great **Scotland Yard**, and its officers were issued with their distinctive **helmet** in 1863 and shrill **whistle** in 1884. Though all police forces had their sleuths, the French ex-con Eugène François Vidocq (1775–1857) is reckoned to be the first **modern detective**. **Fingerprint ID** began with the work of Sir William Herschel (UK) in India in 1858, and the first criminal was identified (using the Bertillon System, 1888, France) by their fingerprints in 1892 (Argentina). Colin Pitchfork was the first criminal to

A nineteenth-century Scotland Yard officer

be convicted after **identification by his DNA** (1988, UK). France had the first **mounted police** (*c.* 1700), the US the first **police car, which was electric** (1899, Ohio) and **police car with a radio** (1928, Detroit). **Interpol** was set up in Vienna, Austria, in 1923.

THE SCENT OF BLOOD

The first attempt by a professional police force to employ dogs in an investigation was a total failure. Frustrated by their inability to catch the notorious Jack the Ripper, London's Metropolitan Police borrowed a pair of bloodhounds, Barnaby and Burgho, and experimented with their scent-following abilities (1889, UK). When the force couldn't decide whether to use the dogs, and failed to pay for their hire, the disgruntled owner took them back. Unfortunately, no one told the investigating officers. When another grisly murder was discovered, the police waited two hours for the arrival of the non-existent bloodhounds before examining the crime scene. Ten years later, the police in the Belgian city of Ghent became the first to deploy dogs in an organized and effective way.

SERVICES

FIREFIGHTING

Fire was an ever-present threat in early cities, and it is no surprise that the first mention of some sort of **firefighting service** comes from Alexandria (Egypt), the world's largest city in the first century BC. Whether or not it used the **pumping mechanism** developed by Ctesibius of Alexandria (third century BC) or Hero of Alexandria (*c.* AD 10 to *c.* 70) remains a moot point. We do know, however, that when Emperor

Augustus established a **military firefighting force** for Rome (the *vigiles*, Italy) in AD 6, he based it on the brigade already operating in Alexandria. Rome's earlier so-called firefighting service, established by the millionaire rogue Crassus, was as much about arson as fire extinguishing. In 1824, Edinburgh (Scotland, UK) established the first **professional, municipal fire brigade** of the modern era. Nine years earlier, the New York slave Molly Williams had become the first known **female firefighter**. **Modern fire engines** may have originated with a pumping machine employed in Augsburg, Germany, in 1518, but the first mobile 'water engine' was built in England by Richard Newsham (1721). A **steam fire engine** capable of projecting two tons of water per minute followed (*c.* 1829, UK), and a **motorized fire engine** was constructed in America in 1905. The portable **fire extinguisher** (other than a bucket of water) was invented by Englishman George Manby in 1819; the first **electric fire alarm** was patented (in error as a 'tire-alarm'!) in 1890 (USA); and the principle of the **smoke detector** was discovered in the 1930s (Switzerland), although the actual device did not reach the market until the next decade.

PUBLIC UTILITIES

The first known **wells** were dug in the Jezreel Valley, Israel, about 8,500 years ago. By 4000 BC the Mesopotamians (Iraq) had installed **sewage systems** with clay pipes, and 1,000 years later the inhabitants of Skara Brae (Scotland, UK) had **running fresh water** and **toilets** (see p.56). **Large-scale urban water supply, drainage and sewage systems** were key features of the cities of the Indus Valley Civilization (*c.* 2500 BC, Pakistan). Because of the vast quantities of ash (or 'dust' – hence 'dustman') produced by its coal fires, London, UK, was the first city with a **waste removal system** (late eighteenth century). On the orders of Eugène Poubelle, in 1884 Paris, France, became the first city to insist that different types of **waste be separated**, some for **recycling**; it may also have been the first major city with a **regular dustbin emptying service**. The first **dustcart** ('steam motor tip-cart') was used in Chiswick, London, (1896–7), and the first **waste incinerator** opened in Nottingham, UK, in 1874. The **wheelie bin** was invented in 1968 (UK). Although large-scale **recycling** did not arrive until the 1990s, **scrap metal** was always valuable, **paper** was being recycled in Japan in 1031, and **rags** in England in 1813. Sweden pioneered recycling **glass** in 1884 and **aluminium cans** in 1982, while Switzerland led the way with **electrical goods** in 1991. Over 2,000 years ago, the Romans built the first **sea wall** at Caesarea Maritima (Sebastos, Israel).

TOLERATION AND EQUALITY

RACE

While some would argue that **racism** is an inherent human characteristic stemming from suspicion of 'the other', its roots have been traced to writers in almost every literate culture who believed their own ethnicity superior to others, including Aristotle (see p.197) and the Arab writer al-Jāhiz (AD 776 to *c.* 869). The coining of the word **ethnocentrism** (1891) to explain such thinking distinguished it from so-called **scientific racism** first observable in the works of Henri de Boulainvilliers (1658–1722, France) and, more significantly, in Carl Linnaeus's division of *Homo sapiens* (see p.6) into five varieties (1767, Sweden). **Racialism** or racism was first used as a **term of disapprobation** in the early twentieth century. The first proposal for a **universal rejection of racism** was made in 1919 (Covenant of the League of Nations) but did not come into being until Article 1 of the UN Charter (1945).

WOMEN'S RIGHTS

In ancient Mesopotamia, Egypt, India and several parts of Africa, women enjoyed much the same rights as men. The **oppression of women** in most Eastern cultures goes back into prehistory, while in Abrahamic cultures (Judaism, Christianity, Islam) it dates from the time of the Old Testament (possibly 1000 BC). The first **proto-feminist** was the Franco-Italian writer Christine de Pizan (1364 to *c.* 1430); Olympe de Gouges (1748–93, France) and Mary Wollstonecraft (1759–97, UK) are generally regarded as the first **modern feminists**. Simone de Beauvoir's *The Second Sex* (1949, France) was the dawn trumpet of the **contemporary feminist movement**. The phrase '**Women's Liberation**' was coined in 1964 (USA). The first **International Women's Year** was 1975, and in 1979 the United Nations adopted the **Convention on the Elimination of All Forms of Discrimination Against Women**.

HOMOSEXUALITY

Research has shown that in 21 per cent of pre-industrial cultures homosexuality was accepted or ignored. In other words, the acceptance of homosexuality, known today as

gay rights, has a very long history. As with women's rights (see opposite page), the struggle against homophobia has focused largely on areas influenced by Abrahamic teachings. In 1791, for example, France became the first country to **decriminalize homosexual acts** between consenting adults. Jóhanna Sigurðardóttir was the first **openly gay head of state** in modern times (2009, Iceland), and Sweden was the first country to allow **transsexuals legally to change their sex** (1972). There is evidence of **same-sex marriages** in ancient Egypt (first millennium BC); the Roman emperor Nero (reigned AD 54–68) separately married two men; Denmark accepted legal **same-sex partnerships** in 1989, and in 2001 the Netherlands gave **same-sex marriages** the same legal status as heterosexual marriages.

WEAPONS

STICKS, SWORDS AND SLINGS

We saw how the first weapons were made many millennia ago (see p.135). After that, the **knife** (one-sided blade) morphed into the **dagger** (two-sided blade), which then stretched into the **sword**, the earliest (in bendy bronze) appearing in Turkey around 5,300 years ago. Iron- and steel-bladed swords were created at the end of the second millennium BC, the **falchion** about 1300 (Europe), the **cutlass** (also Europe) in the seventeenth century; the **rapier** became fashionable in Spain around 1500. **Axes**, **hammers** and **clubs** (see p.135), the forerunners of **maces** and other trauma weapons, date from prehistoric times, though the multi-purpose, Swiss-favoured **halberd** – like the Korean *woldo* and the Chinese

A rubbing depicting Guan Yu (left) and Zhou Cang who wields the Green Dragon Crescent Blade, 1574

guandao – seems to have emerged in the mid-medieval period (tenth to twelfth centuries). The **crossbow** was probably a Chinese invention of at least the seventh century BC. The **sling** emerged in Neolithic times (*c.* 12,000 BC to Bronze Age), and **boomerangs** ('throwing' sticks) even earlier, hurtling through the air in Australia (and Poland – the weapon was not exclusively Australian) as long ago as 30,000 BC.

FIREARMS AND ARTILLERY

The Chinese invented **gunpowder** – probably while seeking out a new medicine – in the ninth century. A century later they were using gunpowder-fuelled 'fire-lances' (**flamethrowers** – see p.218). Loaded with shrapnel, by the twelfth century these had morphed into **cannon** or **bombards**. The earliest hand-cannons (**handguns**) were made in Italy in the late fourteenth century. From these sprang the **arquebus** (fifteenth century, Germany), the **matchlock** (1470s, Germany), the **wheel-lock** (*c.* 1500, Italy and Germany), and **flintlock** (*c.* 1600, France?). The idea of sticking a knife on the end of a gun to form a **bayonet** may have originated, like the name, from the French town of Bayonne in the sixteenth century, or earlier from China. **Rifling** a barrel for greater accuracy emerged in Germany in 1498. **Pistols** first appeared in sixteenth-century Europe (perhaps in France or the Czech Republic). A multi-shot gun with revolving barrels was designed in China (*c.* 1590), and a German gunsmith came up with a single-barrel version with a revolving chamber (the prototype **revolver**) around the same time. The Gatling was the first successful **machine gun** (1862, USA). **Cartridges** started with paper versions (fourteenth century, Europe) before the discovery of **fulminates** (1800, UK) and the manufacture of the **percussion cap** (1807, UK) led to the first **self-contained cartridge** (1808, France), modified versions of which are now universal. **Breech-loading** was devised in fourteenth-century Burgundy (France) but not widely used until the advent of precision engineering in the nineteenth century. Bombards were firing **shells** in mid-fourteenth-century Italy and China, and they were **mounted on wheels** in the 1420s (Czech Republic). The earliest **mortar** fire was apparently heard during the Ottoman siege of Constantinople in 1453. The history of **night-vision sights** began in Hungary with the work of Kálmán Tihanyi (1929), and took practical form in the German army in 1939. Modern **flamethrowers** (*flammenwerfer*) date from 1901 (Germany).

BOMBS AND EXPLOSIVES

We have China to thank for the invention of **bombs**, first in tubes of bamboo (eleventh century), then in metal casing (thirteenth century or earlier). Bombs were dropped from a balloon in 1849 (Austrians attacking Venice, Italy). The first **explosive dropped from an aircraft** was a grenade from an Italian aircraft in 1911; a Bulgarian dropped the first purpose-built bomb the following year, and the first **bomber planes** were the Caproni Ca.30 (Italy) and British Bristol T.B.8 (UK), both in 1913. Naval and land mines originated in medieval China (fourteenth century). Electronic **naval mines** were devised by the Russians (1812), leading to the first **minesweeping** by British warships during the Crimean War (1855). In 1939, German bombers dropped the first **mines by parachute**, and in 1940 they employed the same weapons against land targets. The German army pioneered **anti-personnel mines**, 'silent soldiers' designed to mutilate rather than kill, in 1939. The earliest self-propelled **torpedo** was made in 1866 (UK). The Swede Alfred Nobel invented **dynamite** in 1863, the same year as the German chemist Julius Wilbrand made **TNT** (trinitrotoluene), though its explosive qualities were not seen until 1891. In 1875, Nobel also came up with gelignite, the first **plastic explosive**. **Napalm** was devised in the USA in 1942. While mention is made of noxious or poisonous fumes and smoke in ancient warfare (first millennium BC, China, India, Greece), true **chemical warfare** began in 1914 with the French release of tear gas, prompting the Germans to respond with irritants and then, on 22 April 1915, deadly chlorine gas. Throwing the bodies of dead humans or animals over the walls of besieged towns or castles in order to spread disease (there's documentary evidence from 1347, Ukraine) was an early example of **biological warfare**; it was first deployed systematically, using modern medical knowledge, by the Japanese in their war with China, 1937–45. **Nuclear weapons** are covered on p.150.

THE IMPOSSIBLE GUN

Legend tells us that the teenage Samuel Colt (1814–62, USA) heard soldiers discussing whether there would ever be a firearm capable of firing multiple times without reloading – and the young American made up his mind to create this 'impossible gun'. Early attempts did not go well and when his first pistol exploded on firing, his father refused him further financial backing. Nothing daunted, Colt toured the US demonstrating the powers of laughing gas (see p.65) and entertaining crowds with waxwork shows backed by fireworks. This earned him enough money – boosted by a $300 loan from a family friend – to resume gun-making. He patented his 'impossible gun' (aka the Colt revolver) in London (UK) in 1835 and the US the following February. Though there were further setbacks and disappointments, the 'impossible' dream came true: by the time of his death Colt had sold almost half a million guns worldwide and was worth an estimated $15 million.

WAR MACHINES

ON LAND

The ancient Assyrians (Iraq/Syria) led the way in siege warfare, developing the **siege tower** and **battering ram** in the ninth century BC. Indians are credited with the first siege **catapults** (early fifth century BC), while the Greeks came up with the **ballista** (firing huge arrows or bolts) at the end of the same century. The Chinese came in on the act with the swinging arm **mangonel** (trebuchet) in the fourth century BC and the **mining of walls** around the same time. **Siege guns** are dealt with on p.217 and **chariots** on p.104; **war elephants** were found in Indian armies by the sixth century BC. Mounted infantry – spear-throwers and archers – formed the first **cavalry** (*c.* ninth century BC, Central Asia/Assyria. Animal-powered, armoured war wagons were used in Bohemia, Czech Republic, (*c.* 1420), though the first true **armoured car** was either the four-wheeled, petrol-driven cycle fitted with an armoured shield and a Maxim gun (1898) or the Simms Motor War Car on a Daimler chassis (1899) both in the UK. The idea

War elephants on a seventeenth-century stone relief, India

of an **APC** (Armoured Personnel Carrier) began with the British Mark IX tank (1918), and the US developed the original armoured **scout cars** in the 1930s. To break the stalemate on World War I's Western Front, the UK and France came up with the **tank** at about the same time: the British Mark I was the first to see action (September 1916, perhaps prematurely), while the revolutionary French Renault FT (featuring a revolving gun turret) was the template for the future. **Warships** are covered on p.122.

IN THE AIR

The first **military use of a flying craft** was when the French Aerostatic Corps deployed a tethered balloon to get a good view of what was going on in the Battle of Fleurus (1794). The US Army Signal Corps' Wright Model A was the first **military aeroplane** (1909). The first **aerial combat** took place between Serbian and Austro-Hungarian pilots at the start of World War I (1914), after which Serbia and Austria-Hungary became the first powers to **arm their military planes**. (For the first **bomber**, see p.218) Britain's Royal Air Force (RAF) was the first **air force independent of the army and navy**. Germany produced a prototype **jet fighter** (the Messerschmitt Me 262 Schwalbe, 1942) and **jet bomber** (the Arado Ar 234 Blitz, 1944). The Hawker Siddeley Harrier was the first successful fixed-wing **military VTOL aircraft** (1960s, UK). After the first **war rockets** (see p.128), the Kingdom of Mysore (India) made iron-cased rockets towards the end of the eighteenth century. Further developments in rocket science (see p.128) led to the German V1 pilotless **flying bomb, V-2 guided missile** (1944), and the USSR's R-7, the world's first **ICBM** (intercontinental ballistic missile) in 1957.

Messerschmitt Me 262 Schwalbe, c. 1945

WARFARE

CONFLICT

There is evidence of a **battle** of some sort fought at Jebel Sahaba, Sudan, about 13,000 years ago, though the Battle of Megiddo (fifteenth century BC, Syria) is the first of which we have a reliable record. The first **war** was fought in Mesopotamia (Iraq) around 4,700 years ago between Sumer and Elam, and we may call the rebellions against Sargon the Great of Akkad (*c.* 2350 BC, Iraq) the first **civil war**. World War I was, self-evidently, the first **world war** (1914–18). **Amphibious warfare** began with attacks on ancient Egypt by the mysterious Sea Peoples *c.* 1276–1178 BC; the earliest known **sea battle** was fought between Hittites and Cypriots around 1210 BC (Mediterranean), and the first **air battles** took place at the beginning of World War I (see p.221). The efficacy of **land and air power acting in concert** was demonstrated for the first time during the Battle of Amiens (1918, France), when some 2,000 planes supported an Allied attack of 75,000 men and over 500 tanks. The Battle of the Coral Sea, twenty-four years later, was arguably the first major **air–sea battle**. *The Art of War*, attributed to Sun Tzu, was the first **book on military strategy and tactics** (mid-first millennium BC, China), while at about the same time the Spartans of ancient Greece were pioneering **systematic military training**. Over 1,000 years earlier, Emperor Hammurabi of Babylon (see p.142) had introduced **conscription**.

EARLY DEFENCES

The first **defensive structure** was probably an earth rampart that has long since disappeared, leaving the stone **walls** of Jericho (see p.159) the oldest we know about. The first **castle** was noted on p.159. Earthen **frontier walls** were raised in China from the eighth century BC, though stretches of the Great Wall raised by the First Emperor (reigned 220–210 BC) were the first of lasting significance. **Trench defences** were first devised to halt chariot attacks *c.* 2500 BC (see p.104), and one of the first **moats** surrounded the Egyptian fortress of Buhen (*c.* 1860 BC), requiring some of the first **drawbridges**. The site also featured early **loopholes** and **battlements** (crenellated walls shielding wall-top walkways), though the **portcullis** came later (208 BC, ancient Rome, Italy). Personal defence depended on **armour** (see p.43), the **shield** (an early depiction is Egyptian, *c.* 1300 BC), and the **helmet** (there are early images on the Standard of Ur, see p.104).

THE AGE OF BRICK AND CONCRETE

Artillery gave rise to the small fort known as a **blockhouse**, one of the first of which was Cow Tower, Norwich (1398, UK). The earliest reference to a **pillbox** is 1917 (UK). As firepower increased, targets were safer underground in a **dugout** or **bunker**, terms first widely employed in World War I (1914–18, Europe). The deadly effects of aerial bombardment seen in the Spanish Civil War (1936–9) led to the construction of purpose-built **air-raid shelters** in Spain and elsewhere, and the Cold War produced **fallout shelters** in the 1950s (USA and elsewhere). **Sirens** were invented *c.* 1799, used to alert the fire brigade *c.* 1900, and first warned of an air raid in 1939 (all UK). Meanwhile, an **anti-aircraft gun**, the German *Ballonabwehrkanone* (Balloon Defence Cannon), made its appearance in 1870, as did the military **searchlight** (also German). **Tracer bullets** were a British invention of 1915. The US developed the first **missile defence system** in the 1950s, although the Russians achieved the first **successful ballistic missile intercept** (just testing!) in 1961.

PART VII: CULTURE AND SPORT

SCULPTURE

If the Venus of Berekhat Ram (Israel) is actually a carving and not simply a shapely stone, then, at perhaps half a million years old, it would be undoubtedly the first piece of known **statuary**. Far less controversial is the first clearly **figurative sculpture**, the Lion-man of the Hohlenstein-Stadel (Germany), said to have been made from a mammoth tusk up to 40,000 years ago. The earliest unmistakeable **representation of a human form**, the German Venus of Hohle Fels, also carved out of mammoth ivory, was created a short time afterwards; the first known **stone statue** – the 11.1 cm / 4.4 in. Venus of Willendorf (Austria) – was carved some 10,000 years later. **Lost-wax casting** started *c.* 4000 BC in the Indus Valley (Pakistan), where the earliest **bronze statue** of a human figure (the Dancing Girl of Mohenjo-daro) was made (*c.* 2500 BC). The earliest terracotta figures have the same provenance (3000 BC). **Copper statues** were made in Iraq (*c.* 2600 BC), and the Greeks fashioned **marble statues** and **realistic, life-size human figures** from the sixth century BC onwards, while *Eros* in London's Piccadilly Circus pioneered **aluminium statuary** (1893, UK). The first known **relief**

carving was the Venus of Laussel (*c.* 23,000 BC, France). The Greek Ageladas (late sixth century BC) may have been the first **professional sculptor**, and his workshop the first **school for sculptors**. A competition for the doors of the Florence Baptistery (1403) inaugurated **Renaissance sculpture**. Tiptoeing into very uncertain territory, some might say Constantin Brâncuşi's *Portrait of Mademoiselle Pogany* (1912) introduced **modern abstract sculpture**, and **installation art/sculpture** began with the creations of Marcel Duchamp (1887–1968) and Kurt Schwitters (1887–1948). Finally, on much safer ground, the earliest known **mosaics** were pieced together in Mesopotamia (*c.* 2500 BC).

The Lion-man of Hohlenstein-Stadel, the first known piece of figurative sculpture, *c.* 32000 BC

PAINTING

COLOURS AND BRUSHES

The dawn of painting is a murky and controversial affair. The earliest glimmer may have been the red lines **drawn** on a piece of South African stone, perhaps 73,000 years ago; or the bull **painted** on the wall of the Lubang Jeriji Saléh cave in Indonesian Borneo, (*c.* 37900 BC). This is also the site of the first known **hand stencils**. France's Lascaux cave has clear representations of **human figures** from around 15000 BC, though those in the caves of Brazil's Serra da Capivara National Park and Spain's Cave of Altamira may be older. We are told that the ancient Greeks were the earliest painters of realistic **portraits** and came up with **trompe l'oeil** (both *c.* 750 BC). The first **paint** was ochre based, mixed with fat, egg yolk or water; other colours were made with different minerals. The earliest **tempera** painting is found in Egypt, from roughly the same time as early **frescoes** were being painted in Greece (*c.* 1600 BC). **Watercolour** as a specific art form began in fifteenth-century Europe, and **oil paintings**, using walnut and poppy oils, were created in Afghanistan (*c.* AD 650). **Acrylic paint** was applied in 1940s Germany, no doubt supplied in a collapsible **paint tube**, invented by John Rand in 1841 (USA). The Chinese general Meng Tian is credited with having made the first **paintbrush** (*c.* 300 BC).

STYLES AND SHOWS

As artistic movements have no specific beginnings (or endings), we can get no nearer than the mid-fourteenth century for **Renaissance** painting and the mid-nineteenth for **Impressionist** art. **Cubism** emerged in the first decade of the twentieth century, and Western **abstract art** (long after pioneering splashed-ink work by the eighth-century Chinese artist Wang Mo) in the following decade. **Surrealism** followed in the 1920s. Artists teaching their skills to others – **art education** – is as old as art itself, while painters' workshops were prototype **art schools**. Cosimo de' Medici established the first **academy of art** in 1563 (Florence, Italy). With certain limitations, private **art collections have often been open to public view** (e.g. Rome's Capitoline Museums, 1471), but not until the seventeenth century do we get privately funded **collections open to all** (the first being Oxford's Ashmolean Museum, 1683). We have to wait until the next century before we find a **publicly funded gallery** – the British Museum, 1753. The Paris Salon of 1667 may have been the earliest **art exhibition**. **Art auctions** began with the dispersal of the Earl of Oxford's collection in 1742 (UK), and the first **million-pound** (and **million-dollar**) **painting** was Velázquez's *Portrait of Juan de Pareja*, which tripled the previous record to fetch £2,310,000 (1970, London, UK).

MUSIC

EARLY MUSIC

One might argue that the first **music** was not human at all, but made by birds and other animals long before prehistoric *Homo sapiens* started tapping, clicking, humming and whistling. Judging by the age of the first **musical instruments** (see p.231), **human music** almost certainly came into being at the same time as the first human – in other words, we are musical creatures. Though instruments and illustrations confirm that we have been making music for millennia, we don't know what it sounded like until it was **written down** (*c.* 1400 BC, the Hurrian songs, Syria; some say the Indian musical notation system is even older). The earliest **complete composition** was the Seikilos epitaph (first to second century AD, Turkey). The first known **choral repertoire** belonged to the dramatic chorus in Greek drama (*c.* 700 BC), and **polyphony** – as well as emerging instinctively in Africa – was documented as *organum*, a development of **Gregorian Chant** (AD 600, Italy), in Europe in the second half of the ninth century AD. The first record of a **scale** – the octave scale – is attributed to the ancient Greek mathematician Pythagoras (*c.* 540 BC), while **solmization** – giving each musical note a syllable – began with Guido of Arezzo (*c.* AD 991 to *c.* 1033, Italy), who gave us what in English is *do*, *re*, *mi*, *fa*, *sol*, *la* and *ti*.

MUSICAL MILESTONES

While space precludes listing the first of every musical genre, subgenre and ever-growing sub-subgenre (in January 2019, for instance, there were 343 different types of electronic music), a selection of major firsts might begin with the first **musician/poet** known to us by name, Enheduanna (*c.* 2285 to *c.* 2250 BC), the Sumerian high priestess (Iraq). The first **Chinese Canjun Operas** were performed in the Later Zhao Dynasty (AD 319–351). The Byzantine abbess Kassia (born *c.* 810, Turkey) was perhaps the first **composer known by name whose music still survives** – and certainly the first named **female composer** whose work we can hear. **Western opera** began in Italy (1598), where the **overture** was devised a couple of years later, Arcangelo Corelli (1653–1713, Italy) introduced the **concerto**, and the first **symphonies** were played in Italy in the 1730s and the Austrian composer Joseph Haydn composed the first **string quartet** in the 1750s. Guillaume-Alexis Paris was the first conductor known to use a baton (1794, Belgium).

The Black Crook (1866, USA), is generally accepted as the first **musical**. Singing the **blues** originated among African Americans in the southern United States in the 1870s. 'I Got the Blues' (1908) was the first published piece. The word '**jazz**' was first used in 1915 (USA); the music itself was a fusion of West African and European styles, via **ragtime** (first ragtime piece published 1895/6, USA). Back in Europe, Arnold Schoenberg's *The Book of the Hanging Gardens* had brought **dissonance** into classical music (1908–9, Austria).

Country music grew out of the culture of western Europe with its distinct American tone beginning at a 1927 recording session in Tenessee (USA). Though the term '**pop song**' goes back to 1926, '**pop music**' did not appear until the 1950s (UK). The term '**disc jockey**' was heard in 1935 and appeared in print six years later (USA). Five years later, the American paper *Billboard* published the first **chart** of record sales. The term **rock 'n' roll** was first used in 1951/2 (USA), while **reggae** music was born in Jamaica in the late 1960s. Moving on to more recent developments, while the roots of **rap** go back to African music of long ago, the term did not refer to rhythmic speaking on records until 1971 (USA). Around the same time, inner-city African Americans created **hip-hop** music alongside New York's hip-hop culture.

Folk music is as old as music itself, stretching back to labourers chanting as they went about their tasks; periodically revived and reinvented, it was recognized as a seperate genre in the 1959 Grammy Awards (USA).

The **Motown** sound came from Detroit, USA (the motor town) in 1960, the phrase 'world music' was coined in London in 1987, the band Nirvana launched the **grunge** movement in 1991 (USA), and **garage music** started in the UK *c.* 1995.

FATAL BEAT

Before taking up the lightweight baton, conductors beat time by banging a large wooden staff on the floor beside them. The first conducting fatality occurred in 1687, when the distinguished French musician Jean-Baptiste Lully – no doubt overcome by the momentousness of the occasion – accidentally thumped his staff on his toe while conducting a Te Deum to celebrate King Louis XIV's recovery from illness. The consequent abscess turned gangrenous and within two months Lully was dead.

MUSICAL INSTRUMENTS

Percussion

The first musical instrument – if such it is – was the human **voice**. Then came percussion instruments, beginning with the **drum** (the first known example is said to date from 5500 BC, China). **Snare drums** appeared about the same time as **drumsticks** (fourteenth century, Europe). **Cymbals** clashed in China or the Middle East around 1100 BC; **gongs** resounded in China some 300 years later; triangles tinkled in sixteenth-century England; and **bells** – said to have rung in China 4,000 years ago (roughly when the wooden **xylophone** was made) – did not feature in an orchestra before the eighteenth century (Germany). The German-born composer George Frideric Handel scored a part for the **keyboard glockenspiel** in 1739, and the earliest **tubular bells** were French (1860s).

Woodwind

The **flute** was the first instrument to play notes, and the earliest known is said to be over 40,000 years old (Germany). **Recorders** date from the thirteenth century (Germany), **piccolos** from Italy around 1710, and modern flutes from 1832. Though the ancient Egyptians had a clarinet-type instrument called a *zummara* (*c.* 2700 BC), the first true **clarinet** was made by Johann Denner in 1690 (Germany). The **oboe** (aka **hautbois**) made its orchestral debut in 1657, and the modern **bassoon** later in the same century (both France). The Belgian musician Adolphe Sax patented his **saxophone** in 1846. The wail of the **bagpipes** was first heard not in Scotland but in Turkey around 1000 BC.

Brass

Modern brass instruments such as the **cornet** (early nineteenth century, France), **French horn** (1705; valved 1814, Germany), **bugle** (1758, Germany) and **trumpet** (*c.* 1500 BC, China and Egypt; valved 1818, Germany) all developed from hollowed-out animal horns. The **trombone** (aka **sackbut**) was made in the Netherlands in the fifteenth century, the **serpent** in France in 1590, and the **tuba** in Germany in 1835.

Strings

The Babylonians (Iran) were playing a form of **lute** over 5,000 years ago, and early images of **harps** – developed from hunting bows – are 3,200 years old (Egypt). Next came either the **dulcimer** (*c.* 1500 BC, Iran), the Sri Lankan **ravanahatha** (date unknown) or the Greek **lyre** (1400 BC). The Mongolian **chuurqin**, ancestor of the **morin khuur**, may have been the first instrument played with a **bow** (seventh century), while it is thought that Arabs had developed the **rubab** (seventh century AD, Afghanistan) into an early form of the **violin** by the ninth century. A recognizable three-string violin had emerged in Italy by the 1530s, where the image of a four-string instrument (like most modern ones) was drawn in 1556. **Violas** were being played in Italy in the previous century. Italy was also the home of the **cello** (1535–6) and **double bass** (as a bass **viola da gamba**, 1542). The **electric double bass** was an American invention of 1924. Some form of **guitar** has been around for approximately 3,300 years (Mesopotamia, Iraq), although an electric version was not patented until 1931 (USA) and made its debut in 1936.

Keyboard

The **organ** was a Greek invention of the third century BC. Late medieval and early modern Europe produced a series of new keyboard instruments: the **clavichord** (fourteenth century, Germany), the **harpsichord** (fourteenth to fifteenth century, Germany or Italy), the **virginals** (*c.* 1460, first mentioned in the Czech Republic) and the **spinet** (1631, first mentioned by an Italian). The **piano**, the queen of all keyboard instruments, was invented by the Italian Bartolomeo Cristofori (1655–1731). **Hammond organs** became available in 1935, and Wurlitzer **electric pianos** in 1955 (both USA). The American Elisha Gray made an **electric synthesizer** in 1876.

THEATRE

THE ORIGINS OF THEATRE

There is evidence of **religious theatrical displays** and dramas in ancient Egypt as far back as 2000 BC, of staged **musical shows** in early China (*c.* 1500 BC), and of Sanskrit drama in India from around 600 BC. **Theatre** as we know it – players performing a scripted show on a stage before an audience – emerged in ancient Athens (Greece) during the sixth century BC, perhaps with **tragedy** in 534 BC (the date of the city's inaugural **drama competition**) and **comedy** in 425 BC. The first **theatres** were constructed at roughly the same time. The Greek writer Aeschylus (*c.* 525 to *c.* 455 BC) was the first known **male playwright**, while Hrotsvitha of Gandersheim (*c.* AD 935 to *c.* 1005, Germany) is said to be the first known **female playwright**. When the Greek player Thespis (hence 'thespian') stepped onto the stage in *c.* 534 BC, he became the earliest known **professional actor**. China did not share the West's scruples about women on stage, and it was there, certainly by the Tang dynasty (AD 618–907), that the first **female actors** performed in generally chauvinist musical dramas. By then, if not earlier, China had what were probably the first **theatre schools**. **Puppet theatre**, too, started in China during the Han dynasty (206 BC to AD 220). In 1551 a **commedia dell'arte** performance was recorded in Italy; in the same decade, also in Italy, plays were presented behind a **proscenium arch** for the first time.

Characters in a Chinese puppet theatre, *c.* 1780

STAGE AND CIRCUS

A form of **revolving stage** was pioneered in Japan in the 1750s, though the earliest modern version began turning in Germany in 1896. Candle-lit **indoor theatres** were built in France and Italy at the end of the sixteenth century. Theatre **gas lighting** arrived in the first decade of the nineteenth century, about the same time as **arc lights** (see p.157), followed by **limelight** in the 1820s. **Revues**, descended from street entertainment, are said to date from the opening of the Théâtre des Folies-Marigny in Paris in 1848, though **nude** theatrical performances took place long ago (e.g. in ancient Egypt). New York's Webster Hall is said to be the first **nightclub** (1886, USA), while Berlin hosted the world's

Astley's circus, the world's first, *c.* 1808

first discotheque (1959, Germany). The idea of a **circus** was the brainchild of the retired English cavalryman Philip Astley in 1770. The shows started to be called 'circuses' in 1782 (UK); the 'big top' **tent** was introduced in 1826 (USA); and a **menagerie** was added shortly afterwards (USA).

THE WRITTEN WORD

WRITING

Writing had a long and controversial gestation in two, three or possibly four independent orthographic wombs: certainly Sumer (Iraq) and Meso-America, and perhaps also Egypt and China. Of these, the Sumerian was the first to give birth. First came **tokens** for counting (*c.* 8000 BC), then **pictographs** of these tokens made on clay tablets (*c.* 3500 BC), followed by **phonetic glyphs** (*c.* 3000 BC – a major step forward as it introduced the idea that writing emulated speech), and finally, around 1500 BC, an **alphabet** of letters representing the sounds of speech. The so-called Phoenician alphabet is the oldest. **Egyptian writing** (hieroglyphs) began around 3100 BC, **Chinese** sign-writing *c.* 1200 BC, and **Meso-American** sign-writing *c.* 300 BC. The alphabet that spawned the writing systems of the Middle East and most of Asia, excluding China, was first used in the eighth century BC. The Greek alphabet of about the same time – from which the script you are now reading has evolved – is considered the first **true alphabet** as it gave equal weight to vowels and consonants. To enable Napoleon's soldiers to communicate silently in the dark, the Frenchman Charles Barbier invented a fiendishly complicated **tactile code** that the fifteen-year-old Louis **Braille** simplified into the tactile writing system of today (1824, France).

PEN AND PAPER

The earliest known **writing implement** was the stylus for marking tablets of wet clay (see p.206), followed by the reed brushes and **pens** favoured by the ancient Egyptians (*c.* 3000 BC). Their scribes were writing in **ink** some 400 years later. **Quill pens** were probably made in the Middle East about 100 BC, and the Romans were scratching away with **metal nibs** during the first century AD. Mass production of **steel-nibbed dip pens** started in 1822 (UK), with the **fountain pen** appearing just five years after this (Romania/France). Though a **ballpoint** pen was patented in 1888 (USA), the implement did not become popular until the version made by László **Bíró** in 1938 (Hungary). Several decades before this, Lee Newman had patented a **felt-tip** marker pen (1910, USA). Japan produced the first **fibre-tip** (1962). The **pencil** began with a piece of graphite (1564 onwards, UK), which was then wrapped in wood *c.* 1560 (Italy), and mixed with clay to make the modern **'lead' pencil** (1790, Austria). Before the Chinese invented **paper** (see p.10), writing was done on **papyrus** (fourth millennium BC, Egypt) or **parchment** (third millennium BC, Egypt). To complete our desktop, we need **paper clips** (1867, USA), **bulldog clips** (1944, UK), **clear sticky tape** (1930, USA) and **correction fluid** (1951, USA). Typewriters, carbon paper and word processors are covered on p.138.

BOOKS AND PRINTING

Two Sumerian works share first place in the early **literature** pantheon: a **poem** known as the 'Kesh Temple Hymn' and the *Instructions of Shuruppak*, a **non-fiction** work of proverb -style advice (*c.* 2500 BC). **Scrolls** were written on parchment (see p.10), and the first **books** were Indian palm-leaf manuscripts of the fifth century BC. The codex (many-paged) **bound book** was a Roman invention of the first century AD.

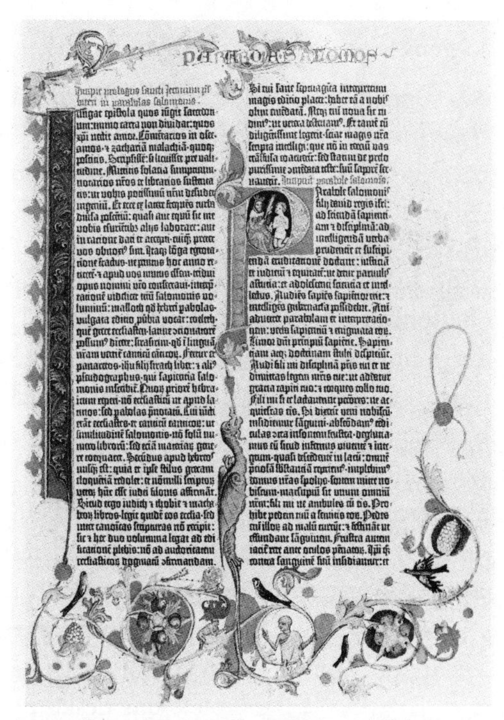

A page from the Gutenberg Bible, *c.* 1454

The earliest known **printed text** is Chinese (*c.* AD 868), while the earliest known **printed book** is *Jikji*, a Korean Buddhist document from 1377. Europe followed later, but more influentially, with Johannes Gutenberg's **printing press** (*c.* 1439); the process was speeded up immeasurably with Richard Hoe's steam-powered **rotary press** (1843, USA). In the meantime, **colour printing** had begun in China *c.* 1346. An early **dot matrix printer** was patented in Germany in 1929; the commercial **laser printer** was launched in 1976; and a **colour laser printer** *c.* 1995 (both USA). A patent for a **3D printer** had been taken out nine years earlier (1986, USA).

LIBRARIES AND GENRES

Written texts needed storing – hence the first **libraries** were Sumerian (see p.193). The first **national library** was in the British Museum (see p.228), and the first **free public library** was in the town of Peterborough, New Hampshire, USA (1833). As in so many aspects of literature, Sumer leads with the earliest **dictionary** (*c.* 2300 BC); much later, the Greek scholar Philo of Byblos (*c.* AD 64–141) wrote the first work that might be considered a **thesaurus**. Before he was killed by the eruption of Mount Vesuvius in AD 79, the Roman author Pliny the Elder was editing his thirty-seven-book *Naturalis Historia*, a work often regarded as the prototype **encyclopedia** (Italy). The first **biography**, which did not become a distinct literary genre until the eighteenth century, is tricky. While some might cite Bible stories (first millennium BC, Israel), most would argue that Giorgio Vasari's *Lives of the Artists* (1550, Italy) marks the true starting point for modern (i.e. non-hagiographic) biography. **Autobiography** began either with the personal section of Sima Qian's *Shiji* (second century BC, China) or with St Augustine's *Confessions* (*c.* AD 400, Algeria). John Newbery's *A Little Pretty Pocket-Book* (1744, UK) may have been the first book written specifically for **children**.

NOVELS AND STORIES

Was the first **novel** the Latin *Satyricon* of Petronius (*c.* AD 50, Italy), the Sanskrit *Daśakumāracarita* of Daṇḍin (sixth to seventh century AD, India), the Japanese *Tale of Genji* of Murasaki Shikibu (eleventh century AD – certainly a good bet for the first novel by a woman), or Cervantes's *Don Quixote*

The early crime fiction story 'Three Apples' appeared in *One Thousand and One Nights*

(1605, Spain)? 'The Three Apples' in *One Thousand and One Nights* (eighth century AD onwards, Arabia, India, Iran) may well predate the first modern **whodunnit** (*The Moonstone*, Wilkie Collins, 1868, UK) by more than a millennium. The same applies to **historical novels**, where Walter Scott's *Waverley* (1814, UK) appeared hundreds of years after Shi Nai'an's *Water Margin* (fourteenth century, China). Though ghosts and spirits feature in much early literature, we might thank Roman playwrights such as Plautus (e.g. his comedy *Mostellaria* – aka *The Haunted House* – *c.* 200 BC) for the

ghost story, while Horace Walpole's *The Castle of Otranto* (1764, UK) is thought to be the earliest **gothic novel**. On the presumption that one can't have true **science fiction** in a pre-scientific age, sci-fi begins with Johannes Kepler's *Somnium* (1608, Germany). *The Swiss Family Robinson* (1812) by the Swiss pastor Johann Wyss is sometimes given credit for launching the **young adult novel**.

WIKI

The wiki (a collaboratively modified website) began with Ward Cunningham's WikiWikiWeb (1994–5). Then came Jimmy Wales's web-based encyclopedia Nupedia (1999), which morphed into the first multilingual, free encyclopedia, Wikipedia, in 2001 (all USA) – quite often the initial source (then double checked, of course!) for this work.

NOVELTIES, NEWSPAPERS AND PRIZES

The monk Matthew Paris made a book with **moveable parts** *c.* 1240, while the *Daily Express Children's Annual* of 1929 was the first book with **pop-ups** (both UK). **Manga** goes back to twelfth-century Japan, with the word first appearing in 1798. American **comics** started with *The Adventures of Mr. Obadiah Oldbuck* (1842), a work originally published in Switzerland in 1827. In 1938 we witnessed the arrival of Superman, the first **superhero** (USA) and the phrase '**graphic novel**' was coined in 1964 (USA). *Relation aller Fürnemmen und gedenckwürdigen Historien* (Germany) was the earliest

A detail from the *Chōjū-jinbutsu-giga*, said to be the first Manga, c. 1200s, Kyoto, Japan

non-government **newspaper** (1605), and a century later *The Daily Courant* became the first **daily paper** (1702, UK). Newspaper **sudokus** were first published in France in 1892–5, and **crosswords** in the USA in 1913. The **Nobel Prize in Literature** was set up in 1901 (Sweden), with other major prizes following: the **Prix Goncourt** (1903, France), the **Pulitzer** Prize (1917, USA), the **Georg Büchner** Prize (1923, Germany), the **Booker–McConnell** Prize (1969, UK; from 2004 the **Man Booker International** Prize), the **Miguel de Cervantes** Prize (1976, Spain), the **Lao She Literary Award** (1999, China), and the **Wole Soyinka Prize for Literature in Africa** (2005).

IT'S IN THE POST

Written message-carrying started with **couriers** in pharaonic Egypt *c.* 2400 BC, and developed into a full-blown government **postal service** either in Assyria *c.* 1700 BC (Iraq/Syria) or, more credibly, in the Persia of Cyrus the Great (559 to *c.* 530 BC, Iran). Collection points were the first **post offices**.

India had a postal service in the third century BC and China by the second century BC. The **public postal service** established by Portugal's Manuel I was arguably the first of modern times (1520). **Letters** appeared shortly after the invention of paper (see p.10) but the earliest known paper **envelope** was Swiss (1615). **Envelopes with windows** were an American innovation of 1902. **Prepaid postage** may date from 1680, but the **adhesive postage stamp** for a **universal, one-price postal system** had to wait until 1840 (all UK). The **pillar** (post) **box** followed in 1849 (Belgium), the plain **postcard** in 1861 (USA), the **picture postcard** in 1872 (Switzerland), and **postcodes** in 1944 (Germany). The first **Christmas card** was sent to King James I and VI in 1611 (UK).

For other means of communication, from drums to emails, see pp.167 and 172.

EDUCATION

LEARNING TO READ AND WRITE

The invention of writing systems (see p.237) required the new skill to be passed on to successive generations; it also led to the growth of a body of significant legal, religious and administrative knowledge. Writing thus led directly to the first **schools**, said to have been those of ancient Sumer (Iraq) or ancient Egypt at the time of Mentuhotep II and his chancellor Khety (*c.* 2020 BC). Before misogynistic religions took hold of education, it seems that boys and girls from well-to-do families were educated alongside each other. Evidence suggests this was the case in Sumer and India during the Vedic era (*c.* 1500 to *c.* 600 BC). In other words, the first **female schooling** occurred at precisely the same time as male. The Jews of the Roman Empire were said to have insisted that **all their children** be educated (first century AD), and 2000 years ago some wealthy Roman families sent their daughters to what may have been the earliest **all-girl primary schools** (Italy).

EXAMS AND TESTING

Our obsession with testing can be traced back to AD 605, when, during the reign of Emperor Yang of the short-lived Sui dynasty, China instituted the first **nationwide exam** as a passport to a government career. Britain copied the idea for entry into the **civil service** (1855), and the US adopted something similar in 1883. Napoleon initiated the French **baccalauréat** in 1808, and the **International Baccalaureate**, based in Geneva, Switzerland, was set up in 1968. **PISA** (Programme for International Student Assessment), a system for ranking countries by educational attainment, published its first results in 2000. Depending on one's definition, **compulsory education** began either in ancient Sparta (ninth century BC onwards) or under the fifteenth-century Aztec Triple Alliance. To give all citizens access to the Bible, the German duchy Palatinate-Zweibrücken was the first territory to instigate **compulsory education for boys and girls** (1592). Twenty-four years later, and for similar reasons, Presbyterian Scotland followed suit, becoming the first nation to require its citizens to **fund** the schooling of the next generation. William Sharp may have been the first teacher to focus exclusively on **science** (first half of nineteenth century, UK).

UNIVERSITIES

Some claim the Moroccan University of Al Quaraouiyine, founded in AD 859 by Fatima al-Fihri, to be the **world's first university**; others – citing academic freedom as a key principle of a true university – prefer the claim of the University of Bologna, going strong since about 1155. The University of Paris awarded the **first doctorate** later in the twelfth century, but no woman was afforded this honour until the University of Avignon made the Spanish Dominican nun Juliana Morell a Doctor of Laws in 1608. She was the **first woman** to receive a university degree of any kind. Medieval Italian universities (see Bologna, above) featured what might be termed the first **graduate schools**, and the College of William and Mary was the first **liberal arts college** (1693, USA). The eccentric and increasingly abused British **degree classification system** (1st, 2i, 2ii, 3rd, Pass) was introduced in 1918.

SPECIALIST EDUCATION

For most of history, training in the military and the creative arts was carried out under the supervision of experienced practitioners. The first **military** (and **naval**) **school** was the Royal Danish Naval Academy (1701), followed by the earliest **army school**, the Royal Military Academy, Woolwich (1720, UK). The Royal Air Force College at Cranwell was the first **military air academy** (1919, UK). On a more pacific note, the papal choir in Rome was probably the first **music school** (fifth century BC, Italy), with **conservatoires** (originally orphanages for the 'saved' who were given a musical education) established in Italy during the sixteenth century. Institutions offering **agricultural education** are difficult to pin down, but these dates are significant: Edinburgh University founded a pioneering **Chair of Agriculture** in 1790, America's Gardiner Lyceum offered vocational **training for farmers** (among others) in 1822, and England's Royal Agricultural College was established in 1845. As much of the language of ballet is French, it is no surprise to find the first **ballet training school** in Paris (1671). For art and theatre schools, see pp.228 and 234.

TOYS AND GAMES

TOPS TO TAMAGOTCHIS

The first **toys** – simple whistles, carts and animals – were those enjoyed by children living in the Indus Valley over 4,000 years ago. The first **yo-yo** may have spun in ancient Greece about 500 BC, and the Greeks also came up with the **mechanical puzzle** (third century BC), a forerunner of the **Rubik's Cube** (1974, Hungary). **Dolls** made almost 5,000 years ago were found in Egyptian tombs, as were **doll's houses**. However, these were not playthings. The first true child's doll was found on Pantelleria (*c.* 2000 BC, Italy), and doll's houses were not made for children until the eighteenth century (UK). Together with the **rocking horse** (*c.* 1600), there were a number of innovative English toys that included the **jigsaw** (1767), the **kaleidoscope** (1817), the **zoetrope** (1833), **plasticine** (1897), and **Meccano** (1898). Germany marketed a complete **model train set** in 1891, and America had an **electric train set** by 1897. Commercial production of stuffed toys began in Germany (1880), but the USA stole a giant march with the **teddy bear** (1903). A British company made

Hatchimal toys were launched in 2016

the first **self-locking bricks** in 1939, only to be trumped by **Lego** (1949, Denmark) at the same time as **diecast models** were appearing (Japan, USA, Europe). The Frog company had made the first **scale model plastic kits** in 1936. The world was introduced to **Play-Doh** in 1956, and to **Barbie** three years later. The **Tamagotchi** (1996, Japan) began the craze for electronic creatures that led to robotic **Hatchimals** in 2016.

The earliest known **board game** was the Egyptian senet (*c.* 3500 BC). A **draughts**-type game was played in Mesopotamia (Iraq) *c.* 3000 BC, and a game resembling **backgammon**, using the first **dice**, emerged in the same region around 2800 BC. **Jacks** (**knucklebones**) originated in China, with an early reference in the fourth century BC. Around a millennium passed before the first **chess** game was played (sometime between the third and sixth century AD, India), and another before we meet **dominoes** (thirteenth century, China). **Modern board games** began with *A Journey Through Europe* (1750s, UK), preparing us for ***Monopoly*** (1935, USA). Once the ancient Egyptians had made **balls** (*c.* 3200 BC), our distant ancestors invented the games that evolved into the many types of **bowling**, including the French **boules** or **pétanque** (1910). Americans devised the **automatic ten-pin bowling alley** in 1946, and the first electronic **pinball machine** (1933) from **bagatelle**, a game that had evolved in France towards the end of the eighteenth century. A croquet-style game, **billiards**, started outdoors in the 1340s and moved onto an indoor table by the time of King Louis XI (1461–1483, France), but the oldest reference to **croquet** itself comes from 1856 (UK). An Englishman devised the modern **dartboard** (1896) and another patented **table football** in 1923, while the French first played **charades** in the early nineteenth century. **Playing cards** originated in China in the ninth century; the **suits** (spades, hearts, diamonds, clubs) are a fourteenth-century French version of earlier Egyptian symbols, and the Americans added a **joker** around 1860. In 1720 an Italian devised the **gaming wheel**, enabling the French to come up with the modern form of **roulette** in 1796. The first modern gaming house, or **casino**, opened in Venice in 1638 (Italy).

DART DEATH

The first confirmed reference to the game of darts comes from 1819. It was then known as 'Puff and Dart' because the missile was not thrown, as today, but launched by a blowpipe. As careless (or tipsy) players occasionally sucked rather than blew, thereby swallowing the dart with often fatal consequences, this form of the game eventually died out, unsurprisingly.

RELIGION

IN THE BEGINNING

Though some dispute this, the first signs of human beings' **religious instincts** were when they started deliberately **burying their dead**. The trouble is, no one can be certain when this was – anything between 300,000 and 30,000 years ago. We are on safer ground with the first **figures and images of religious significance**, and even safer with **religious architecture** (see p.159). Experts disagree whether writings from Mesopotamia (Iraq) *c.* 2600 BC or those from Egypt *c.* 2400 BC are the first religious **texts.** The origins of most major faiths are shrouded in mystery, but it is possible that the first **Zoroastrians** lived in the second millennium BC, **Hindus** and followers of **Judaism** in the sixth century BC, **Confucians** and **Buddhists** (if Confucianism and Buddhism are religions) in the fifth century BC, **Taoists** in the fourth century BC, **Jains** in the second century BC, **Christians** in the first century AD, **Muslims** in the seventh century, practitioners of **Shinto** in the eighth century, and **Sikhs** in the sixteenth century. The first signs of the **Sunni–Shia division** in Islam occurred in the middle of the seventh century AD, and the initial division of the Christian Church between **Roman Catholic** and **Orthodox** occurred in 1054 (Italy/Turkey), with the first **Protestants** splitting from the Roman Church in the 1520s (Germany).

MANIFESTATIONS AND OPPOSITION

The first religious texts (see p.250) were the earliest **prayers**; the first **hymns** were sung in seventh-century BC Greece; and the first **psalm** (No. 29) is believed to have been written between 445 and 333 BC. The first known **priest/priestess**, known as an En, was the Sumerian woman Enheduanna (*c.* 2285 to *c.* 2250 BC). Christian priests and bishops emerged in the middle to late first century AD, but deciding on the first Muslim priest or **imam** depends on whether one's outlook is Shia or Sunni. **Prophets** are equally tricky. Muslims believe the first prophet was Adam, the first human being; however, as they and Jews and Christians accept Abraham (possibly sixth century BC) as a prophet, he also has a claim to be number one. Reports of the first Hindu **miracles** date from the eighth century BC, Judaic ones from the sixth century BC, Christian from the first century AD, Buddhist from the sixth century, and Muslim (*ayahs*) from the seventh century. A definitive first **martyr** is impossible to ascertain; nevertheless, arguments can be made for both the Greek scholar Socrates (399 BC) and St Stephen (*c.* AD 34), the first (if one excludes Jesus) Christian martyr. **Cremations** were carried out in Australia around 42,000 years ago, and the first known **tomb** (as opposed to a simple grave) was built in the Sinai desert (*c.* 4000 BC, Egypt). **Religious wars** began with monotheism, the first being the Muslim-Arab conquests of AD 622 to *c.* 750. **Atheism** – life without one or more deities – originated among Buddhists, Hindus and Taoists during the sixth century BC, and the word '**agnostic**' was coined by the English scientist and thinker T. H. Huxley in 1869. Religious buildings are covered on p.159.

SPORT

ON YOUR MARKS ...

No doubt nimble-footed Neolithic and Neanderthal kids were racing each other and kicking fir cones between suitably spaced trees many millennia ago, but the first visual evidence of sport comes from cave paintings: **foot races** and **wrestling** (*c.* 13,300 BC, France), **swimming** and **archery** (*c.* 6000 BC, Libya), and **sumo** (first century AD, Japan). Sumer (Iraq) held **boxing** matches in the third millennium BC, and provides the first evidence of **angling** (*c.* 2500 BC). Shortly after this,

A stone carving depicting a boy and his slave playing a game of *episkyros* (believed to have resembled rugby), fifth century BC

we find many sports being enjoyed in ancient Egypt: **horse racing**, **fencing**, **gymnastics**, **weightlifting**, and a wide variety of what we now call **track and field** events, including running, **long jump**, **high jump**, **javelin** and **tug of war** (all *c.* 2000 BC). The first **multi-sport event**, the **Olympic Games** (initially foot races only) took place in 776 BC (Greece). Though the Olympics were soon a male preserve, the first race is said to have been a women-only event, and Greece's four-yearly Heraean Games were the first regular **sports event for women** (possibly *c.* 772 BC). The Chinese probably held the earliest **boat races** (**rowing/canoeing**) around 2,000 years ago, and they were playing *cuju*, a sort of **football**, a couple of centuries before (*c.* 250 BC). The origins of **polo** are obscure, though we know it was played in modern-day Iran *c.* 250 BC. One of the first **ball games** was a mix of **racquets** and **handball** played in Meso-America *c.* 1400 BC. There are references to a sort of **hockey** (*kerētízein*, *c.* 510 BC) and a rough, **rugby**-like game (*episkyros*, *c.* 350 BC) in ancient Greece. **Tennis** and **fives** may go back to twelfth-century France, and British tennis star Tim Henman's great-grandmother, Ellen Mary Stawell-Browne, was the first woman to serve overarm at the Wimbledon Championships (1900, UK); **golf** was banned (and was therefore being played) in Scotland in 1457, and the Dutch claim to have instigated **sailing** as a sport in the seventeenth century.

Notable firsts for more modern sports include the writing of rules for **American football** in 1876, then the foundation of the NFL in 1920, and the inaugural Superbowl in 1967. **Baseball**'s famous World Series started in 1903 (USA), and the rules of **basketball** (USA) appeared in 1891 (men, USA) and 1892 (women). Boxing was first recorded in the original Olympics in 688 BC, with the modern rules established in 1885–7 (UK). The UK also drew up the laws of **cricket** and participated in the first test match (England v Australia, 1877). **Disability sport** began with the first multi-sport games held at Stoke Mandeville Hospital in 1948 (UK), leading to the Paralympic Games of 1960.

The first **equestrian sport** was chariot racing in the original Olympics (684 BC), with the first classic horse race, the St Leger, taking place in 1776 (UK). Modern **football** (**soccer**) dates from the foundation of the Football Association in 1863 (England, UK), with the first world cups held in 1930 (men, Uruguay) and 1991 (women, China). The first rules of **golf** were penned in 1744 (Scotland, UK). Long before this, Mary Queen of Scots is alleged to have been the first female player (1567). Major tournaments began in the UK with the Open (1860, men) and the Curtis Cup (women, 1932). The earliest recorded **judo** training schools were in Japan: 1882 (men) and 1923 (women). The sport's rules were drawn up in 1900, also in Japan.

Leander is the oldest **rowing** club (UK, 1818), and the first recorded race took place ten years later (Oxford v Cambridge). The laws of **rugby union** were produced in Rugby School in 1845 (UK), the Rugby Football Union was the sport's earliest organisation, and England v Scotland the first international match (both UK, 1871). **Rugby league** started with the Northern Rugby Football Union's organisation and laws of 1985 (UK). Finally, the first formal organisation for **water sports** was probably the National Swimming Society (England, UK), which wrote its rules in 1908, though the first swimming races were held (in the sea!) at the 1896 Olympics (men), with women following in 1912.

ROYAL REGATTA

For millennia, sailing was serious commercial business. During the first half of the seventeenth century, however, the Dutch were zooming around in their nifty jaghts (yachts, invented 300 years earlier) for fun. England's exiled king Charles II joined in, and when he returned home in 1660 he ordered his own personal yacht, *Katherine*. The next year, in what is said to be the first yacht race, he challenged his brother's yacht, *Anne*, to a forty-mile race. The king won, and the sport of yachting has never looked back.

WINTER SPORTS

Britain's National Skating Association was the sport of **ice skating**'s earliest organizing body, and its 1.5-mile challenge the first **speed skate** race (both 1879). Then came the International Skating Union (1892) and a World Championship (1893), with the speed and figure skating entering the Winter Olympics in 1924 (1960, women). Competitive **figure skating** reputedly dates from the moment Jackson Haines declared himself America's national champion in 1864. The first World Figure Skating Championships were held in the Netherlands in 1896, and 1902 saw the first woman competitor. It became an Olympic sport in 1908.

Meanwhile, an organized **ice hockey** game had taken place in Canada (1875), and the sport's rules were drawn up within a couple of years. By 1920 (1998, women) the sport was in the Olympics. **Curling** goes back much further, with the first club (the Kilsyth Curling Club – still going strong) founded in Scotland in 1716. The sport made a short-lived Olympic appearance in 1924, but was revived with the formation of the International Curling Federation in 1966.

Norway reported a **skiing** race in 1843 and a ski jumping competition in 1866, but Switzerland hosted the first modern slalom competition in 1921. An International Ski Congress (1910, Norway) paved the way for the establishment of the Fédération Internationale de Ski (FIS) and Olympic acceptance (both 1924).

Competitive **sledding** began in Switzerland in 1883, with an international organization (the International Sled Sports Federation) formed in 1924, the same year as sledding entered the Olympic Games. The **snowboard** was invented in 1965 (USA) and swiftly grew to be a sporting phenomenon: the first competition took place in 1968, the International Snowboarding Federation (ISF) was formed in 1990, and within eight years the sport was in the Olympics. The inaugural **Winter Olympics** were held in 1924, though people had been having fun on the snow for centuries before that: the earliest picture of a **snowman** is dated 1380 (Netherlands).

MOTOR SPORT

A **race** between a pair of **steam engines** took place in 1867 (UK), and the first organized **motor car race** twenty years later (1887, France). The Paris–Bordeaux–Paris event of June 1895 was probably the first **car rally**, though the 1911 **Monte Carlo Rally** is the official prototype of such events. Brooklands in Surrey was the first purpose-built **motor racing track** (1907, UK). The roots of **Grand Prix** racing stretch back to 1906 (France), with **F1** World Championship racing kicking off in

1950 (UK). **Formula E** (electric) was launched in 2014. A **car travelled at 100 mph** for the first time in 1905 (USA/UK). The career of Camille du Gast, the first top-class **woman racing driver**, began in 1901. On retirement, she took up **powerboat** (**speedboat**) racing, which had started in 1903. Two-, three- and four-wheeled vehicles raced together until 1904, when **motorcycle racing** declared its independence with the formation of the Fédération Internationale du Motocyclisme. It organized its first race the following year (France) and the first **Tourist Trophy** (**TT**) in 1907 (British Isles). The first race in the **Sidecar** World Championship took place in 1949 (UK). **Speedway**'s history is shrouded in mystery before the emergence of a World Championship, 1931–6. An American Excelsior was the first **motorcycle to clock 100 mph** (1912).

DEATH ON TOUR

Cycle racing began in Paris in 1868, when the winner rode a wooden bike with iron-rimmed wheels. The first velodrome was constructed in 1877 (UK), and the International Cycling Association (ICA) took charge of rules and championships in 1892. Though the sport has been in the modern Olympics from the start (1896), it is the world-famous Tour de France (1903) that grips the imagination – not always for honourable reasons. The event has been directly or indirectly responsible for the deaths of seven spectators and assistants – and four riders: one died of heart failure, one fell into a ravine, one drowned, and the fourth crashed head first into a rock.

KEEPING UP APPEARANCES

HAIR

For reasons of hygiene and aesthetics, barbers have been cutting, trimming and **shaving** since around 3500 BC, when the earliest known metal **razors** were made (Egypt). The Egyptians also sported the first recorded **wigs**, and are said to have pioneered **waxing** (beeswax) and **plucking** with tweezers. Cutting was done with sharp stones, shells or knives until the invention of **scissors** in Mesopotamia *c.* 1500 BC (Iraq). The first **nail clipper** was patented in 1875 (USA and UK). Meanwhile, the first **straight** (or cut-throat) **razor** had been made in Sheffield (UK) in 1680 and continued in widespread use until the arrival of the **safety razor**: first conceived in 1762 (France); first patented in 1847 (UK/USA); and first called a 'safety razor' in 1880 (USA). A patent for King Camp Gillette's **double-edged safety razor** was granted in 1904. By then, a patent for an **electric razor** had been granted (1898, USA), though a practicable device did not go on sale until 1931 (USA). Gillette's razors featured **disposable blades**, but not until the Bic one-piece polystyrene razor (1975, France) did the whole thing become disposable. While various types of **hair cream** had been applied since ancient times, the first **hairspray** did not go on sale until 1948 (USA).

MAKE-UP

The ancient Egyptians took the lead in the make-up stakes, with **lipstick**, **rouge** and **face cream** making their appearance in the fourth millennium BC. **Eyeliner** (kohl) was applied *c.* 3100 BC and **hair dye** (henna) certainly by 1574 BC. The Chinese came up with an early form of **nail varnish** *c.* 3000 BC. **Tattooing** is a very ancient art, the first example of which was found on Ötzi the Iceman (see p.60) who lived *c.* 3250 BC (Austria/Italy). The first known **perfume** was made in Cyprus at the end of the third millennium BC, and Tapputi, the earliest known **perfume-maker**, lived in Mesopotamia (Iraq) *c.* 1200 BC. **Toilet water** (eau de toilette) was made in fourteenth-century Hungary. **Shampoo** was an Indian invention of many millennia ago, but the liquid form did not go on sale until 1927 (Germany). **Aftershave** has been traced back to adverts for 'Persian (or Naples) soap' (1744, UK). In more recent times, American manufacturers marketed **deodorant** (1888, Mum), **talcum powder** (1894, Johnson & Johnson) and **antiperspirant** (1903, Everdry).

EXERCISE

The first known **fitness regimes** were followed by the ancient Persians, who in the first millennium BC built **gyms**, known as *zurkhaneh* ('houses of strength'). San Francisco's 1912 'Bay to Breakers' launched the **fun run** (USA). Built before the first bicycle, Francis Lowndes's Gymnasticon exercise machine (1796) pioneered the **exercise bike** (UK), and a modern **running machine** (exercise treadmill) went on sale *c.* 1968 (USA). **Rowing machines** date back to ancient Greece of the fourth century BC, with the modern, hydraulic version appearing in 1872 (USA). **Yoga** was first practised in India some time between 5,000 and 3,500 years ago, and Joseph **Pilates** (1883–1967) is remembered in the popular form of physical training that bears his name (Germany). **Personal fitness trackers** may be said to date from the

1895 distance-calculating 'Cyclometer' (USA), advancing to the Manpo-kei **pedometer** (1965, Japan – launching the **10,000-step benchmark**), to wearable **heart rate monitors** (1977, Finland) and the **Fitbit** (2008, USA).

Lowndes's Gymnasticon, the first known exercise machine, 1797

HOLIDAYS AND AMUSEMENTS

TIME OFF

The first **holidays**, celebrated for at least 4,000 years, were those welcoming the arrival of a new year (Babylon/Iraq). Legislation **limiting hours of work** was passed in 1802 (UK); an international convention for **paid holidays** was drawn up in 1936; and Sweden pioneered statutory **parental leave** in 1976. The earliest **hotels** probably took guests in either Persia (Iran) or Greece some 3,000 years ago, while **inns** are supposedly a Roman innovation (late first millennium BC, Italy). Britain played a major role in creating the modern holiday industry with the first **travel agent** (1758), **organized tour** (1841, Thomas Cook), and the **package holiday** abroad (1952, to Mallorca). For the young and vigorous, there were also the British organizations of **Boy Scouts** (1909) and **Girl Guides** (1910). The benefits of **sea bathing** were noted in the early eighteenth century, producing the first **bathing machine** (1735, UK). A tented site on the Isle of Man (1894, British Isles) launched the idea of the **holiday camp**, the first permanent one going up twelve years later (UK). **Amusement parks**, combining the fun of the fair and the pleasure garden, are believed to date from Chicago's World's Columbian Exposition (1893, USA), with a park constructed on a permanent site in 1895 (also USA). **Disneyland**, California (USA), opened in 1955. Amusements started with the **carousel** (eighteenth century, Europe), though there are reports of some sort of slide **roller coaster** in Russia in the previous century. A wheeled roller coaster was opened in Paris in 1817, and George Washington Gale **Ferris** built the first **wheel** that bears his name in 1893 (USA).

GADGETS FOR FUN AND FITNESS

Bone **skates** or gliders were worn in Finland around 2000 BC; true skates, with steel blades that cut into the ice, were a medieval Dutch creation. **Roller skates** were reported to have rumbled across a British stage in 1743, but a patent for the invention was not taken out until 1760 (Belgium). In Polynesia, where ice was unknown, **surfing** was the fun pastime for hundreds of years before entering written records in 1767. From there the sport spread to California, USA,

An unlikely combination – a suited man on a space hopper, c. 1970

where enthusiasts fixed roller skates to the bottom of their boards when the rollers were disappointingly low – and the **skateboard** was born (late 1940s). Hoops of one form or another were among the earliest toys, but not until *c.* 1957 did **hula hoops** become a global craze (beginning in the USA). Throwing the **discus** featured in the original Olympics (776 BC, Greece) and led, via a whirling cake tin, to the plastic **frisbee** of 1948. Twenty years later (1968), an Italian inventor dreamed up the unlikely **space hopper**.

ACKNOWLEDGEMENTS

The author would like to thank Bob Cromwell (https://toilet-guru.com); Colin Brown and Julian Anderson for their generous help in compiling this work; David Inglesfield for his many helpful suggestions and corrections; the book's model editor, the ever-patient Gabriella Nemeth; and Lucy Ross for helping to check the book's 6,000-plus facts and fashion early crude jottings into something resembling coherent prose.

SELECT BIBLIOGRAPHY

C.R. Cheney and Michael Jones, *Handbook of Dates for Students of British History*, CUP, 2008.

Peter D'Epiro, *The Book of Firsts: 150 World-Changing People and Events from Caesar Augustus to the Internet*, Anchor Books, 2010.

Patricia Fara, *Science: A Four Thousand Year History*, OUP, 2010.

Ian Harrison, *The Book of Firsts*, Cassell, 2006.

Adam Hart-Davis, ed., *History: From the Dawn of Civilization to the Present Day*, DK, 2015

John Keegan, *A History Of Warfare*, Pimlico, 2004.

Andrew Marr, *A History of the World*, Pan, 2013.

Richard Overy, *The Times Complete History of the World*, William Collins, 2015.

Philip Parker, *World History: From the Ancient World to the Information Age*, DK, 2017.

Carl Ploetz, Hans Wilhelm Gatzke, William L. Langer, and William L. Langer ed., *An Encyclopedia of World History*, Harrap, 1956.

Roy Porter, *The Greatest Benefit to Mankind: A Medical History of Humanity*, Fontana Press, 1999.

J M Roberts & Odd Arne Westad, *The Penguin History of the World*, Penguin, 2014.

Patrick Robertson, *Robertson's Book of Firsts: Who Did What for the First Time*, Bloomsbury, 2011.

Thomas Stevens, *Around the World on a Bicycle*, with introduction by Nick Crane, Century, 1988.

PICTURE CREDITS

Page 6: Wikimedia Commons / hairymuseummatt (original photo) / DrMikeBaxter (derivative work) / CC BY-SA 2.0.

Page 9: skyfish / Shutterstock.

Page 10: Illustration from *China: Its Costume, Arts, Manufactures, &c.*, Vol. 2, London, 1812.

Page 22: Everett Collection Historical / Alamy Stock Photo.

Page 23: McClure's Magazine Vol. 35, No. 06, October 1910.

Page 24: Illustration from *Traitez nouveaux & curieux du café, du thé et du chocolate*, Philippe Sylvestre Dufour, Lyon, 1685.

Page 29: Anonymous depiction of the interior of a London coffee house, British Museum, London; image courtesy of Wikimedia Commons.

Page 30: © Science Museum / Science & Society Picture Library. All rights reserved.

Page 34: Wikimedia Commons / Victoria and Albert Museum / David Jackson / CC BY-SA 2.0.

Page 39: Wellcome Collection / CC BY 4.0.

Page 42: Illustration from Carson, Pirie, Scott & Co. catalogue, 1893.

Page 43: Illustration from *The Ceremonial Usages of the Chinese, B.C. 1121*, translated by William Raymond Gingell, London, 1852.

Page 48: An East Greek spoon, late Classical to Hellenistic period, c. 4th century B.C. (silver) / Private Collection / photo © Christie's Images / Bridgeman Images.

Page 53: © Science Museum / Science & Society Picture Library. All rights reserved.
Page 54: akg-images / Fototeca Gilardi.
Page 66: Illustration of Doktor Schnabel von Rom from *Die Karikatur und Satire in der Medizin* by Professor Dr. Eugen Hollander, Stuttgart, 1921.
Page 68: Picture of Alexis St. Martin from *Life and Letters of Dr. William Beaumont* by Jesse S. Myer, A. B., M. D., St. Louis, 1912.
Page 74: Illustration of a pharmacy from *Microcosmus hypochondriacus*, Malachias Geiger, Munich, 1651 / Wellcome Collection / CC BY 4.0.
Page 78: Wellcome Collection / CC BY 4.0.
Page 84: Articulated artificial right arm, Europe, 1501-1550 / Science Museum, London / Wellcome Collection / CC BY 4.0.
Page 94: Wellcome Collection / CC BY 4.0.
Page 98: Dr. Sheffield's First Toothpaste – Crème Angélique / Sheffield Phamaceuticals' Private Archives.
Page 104: Standard of Ur, British Museum, London / Wikimedia Commons / Babelstone / LeastCommonAncestor / CC BY-SA 3.0.
Page 107: Mary Evans Picture Library.
Page 113: Photo from *Cassell's Book of Knowledge* by Harold F. B. Wheeler, 1924.
Page 119: Newscom / Alamy Stock Photo.
Page 124: Library of Congress / Prints and Photographs Division / LC-USZ62-110384.
Page 126: Library of Congress / Prints and Photographs Division / LC-USZ66-6166A.
Page 129: Photo courtesy of NASA.

Page 130: Photo courtesy of NASA.
Page 137: Illustration by Peter Dunn, © Historic England Archive (inset); photo Heritage Image Partnership Ltd. / Alamy Stock Photo.
Page 143: Photo courtesy of Ørsted.
Page 147: © Science Museum / Science & Society Picture Library. All Rights Reserved.
Page 154: Billion Photos / Shutterstock.
Page 157: © Science Museum / Science & Society Picture Library. All Rights Reserved.
Page 160: Library of Congress / Prints and Photographs Division / LC-DIG-ppmsca-41005.
Page 168: Photo Tim Boyle / Bloomberg via Getty Images.
Page 170: madmickandmo /iStockphoto
Page 175: PictureLux / The Hollywood Archive / Alamy Stock Photo.
Page 180: INTERFOTO / Alamy Stock Photo.
Page 190: Photo © Fiona Slater.
Page 195: Heritage Image Partnership Ltd. / Alamy Stock Photo.
Page 200: Roger-Viollet / TopFoto.
Page 210: Infinity Images / Alamy Stock Photo.
Page 216: Philadelphia Museum of Art / Gift of Major General and Mrs. William Crozier, 1944.
Page 220: Wikimedia Commons / Dharma / CC BY 2.0.
Page 222: Photo courtesy of The National Museum of the USAF.
Page 226: Heritage Image Partnership Ltd. / Alamy Stock Photo.
Page 235: Deutsches Ledermuseum / Wikimedia Commons / Dr. Meierhofer / CC BY-SA 3.0.

Page 236: Illustration from *Microcosm of London*, 1808 / *57-1633, Houghton Library, Harvard University.

Page 239: Library of Congress / Prints and Photographs Division / LC-USZ62-51844.

Page 241: Illustration from *Stories from the Arabian Nights*, retold by Laurence Housman, drawings by Edmund Dulac, London, 1911.

Page 243: Illustration from the first scroll of the Choju-jinbutsu-giga, Kosan-ji Temple, Kyoto, Japan / Creative Commons / Public Domain.

Page 248: Jarretera / Shutterstock.

Page 252: akg-images.

Page 260: Wellcome Collection / CC BY 4.0.

Page 262: TopFoto.

INDEX